Shrinking Planet
and
Geopolitics of Energy

Shrinking Planet and Geopolitics of Energy

Rakesh Kumar

RANDOM PUBLICATIONS
NEW DELHI (INDIA)

Shrinking Planet and Geopolitics of Energy

ISBN 978-93-5111-500-7

Published in 2015 in India by

RANDOM PUBLICATIONS

4376-A/4B, Gali Murari Lal, Ansari Road
New Delhi-110 002
Phone : +9111-43580356, 011-23289044, 011-43142548
e-mail: sales@randompublications.com,
info@randompublications.com, randomexports@gmail.com

Reprint 2022

Type Setting by : Friends Media, Delhi-110089
Printed at : Replika Press Pvt. Ltd.

Preface

A rapidly changing global energy scene, dominated by volatile oil and gas prices, the emergence of powerful new consumers in the Asia-Pacific region, reserve depletion within the OECD and the instability in the energy producing regions, caused by domestic, regional and international political actions, have contributed to the instability of energy markets. Geopolitics refers to the way geography, politics, and economics influence international relations, individual countries, their foreign policies, as well as their economic and political security.

Energy geopolitics in the 21st century is partially a legacy of how energy has impacted on nations' domestic or foreign policy agendas over the past 50 years. In the current geopolitical environment the political manifestation of energy access and resources may not appear as a first order of magnitude but they are present if one digs beneath the surface. When nations' choose to conveniently ignore or to underestimate the energy quotient in their own national or collective security equation, vulnerabilities emerge and risks increase for becoming embroiled in a destructive tug-of-war over resource access and acquisition.

This book discusses the way that rising powers and a shrinking planet have created a new geopolitics of energy. It emphasizes the increased focus on areas of the world that have not been part of geopolitics in the past. It describes situations in which governments step in to manage the energy procurement and supply of their countries, promoting certain objectives that do not align with those of oil-purchasing countries. The book argues that the amount of oil available on the market is not always enough for developed nations, and that demand will only increase as developing countries continue to need greater supplies, aggravating conflicting interests.

Author

Contents

1

Global Energy Systems

Energy is at the heart of most critical economic, environmental and developmental issues facing the world today. Clean, efficient, affordable and reliable energy services are indispensable for global prosperity. Developing countries in particular need to expand access to reliable and modern energy services if they are to reduce poverty and improve the health of their citizens, while at the same time increasing productivity, enhancing competitiveness and promoting economic growth. Current energy systems are inadequate to meet the needs of the world's poor and are jeopardizing the achievement of the Millennium Development Goals (MDGs). For instance, in the absence of reliable energy services, neither health clinics nor schools can function properly. Access to clean water and sanitation is constrained without effective pumping capacity. Food security is adversely affected, often with devastating impact on vulnerable populations.

Worldwide, approximately 3 billion people rely on traditional biomass for cooking and heating, and about 1.5 billion have no access to electricity. Up to a billion more have access only to unreliable electricity networks. The "energy-poor" suffer the health consequences of inefficient combustion of solid fuels in inadequately ventilated buildings, as well as the economic consequences of insufficient power for productive income-generating activities and for other basic services such as health and education. In particular, women and girls in the developing world are disproportionately affected in this regard.

A well-performing energy system that improves efficient access to modern forms of energy would strengthen the opportunities for the poorest few billion people on the planet to escape the worst impacts of poverty. Such a system is also essential for meeting wider development objectives. Economic growth goes hand in hand with increased access to modern energy services, especially in low- and middle-income countries transitioning through the phase of accelerated industrial development. A World Bank study indicates that countries with underperforming energy systems may lose up to 1-2 per cent of growth potential annually as a result of electric power outages, over-investment in backup electricity generators, energy subsidies and losses, and inefficient use of scarce energy resources.

At the global level, the energy system – supply, transformation, delivery and use – is the dominant contributor to climate change, representing around 60 per cent of total current greenhouse gas (GHG) emissions. Current patterns of energy production and consumption are unsustainable and threaten the environment on both local and global scales. Emissions from the combustion of fossil fuels are major contributors to the unpredictable effects of climate change, and to urban air pollution and acidification of land and water. Reducing the carbon intensity of energy – that is, the amount of carbon emitted per unit of energy consumed – is a key objective in reaching long-term climate goals. As long as the primary energy mix is biased towards fossil fuels, this would be difficult to achieve with currently available fossil fuel-based energy technologies. Given that the world economy is expected to double in size over the next twenty years, the world's consumption of energy will also increase significantly if energy supply, conversion and use continue to be inefficient. Energy system design, providing stronger incentives for reduced GHG emissions in supply and increased end-use efficiency, will therefore be critical for reducing the risk of irreversible, catastrophic climate change.

It is within this context that the UN Secretary-General's Advisory Group on Energy and Climate Change (AGECC) was convened to address the dual challenges of meeting the world's energy needs for development while contributing to a reduction in GHGs. AGECC carried out this task in a rapidly changing environment in which energy was often a key factor: the sensitivity of the global economy to energy price spikes; increased competition for scarce natural resources; and the need to accelerate progress towards achievement of the MDGs. The world's response to climate change will affect each of these issues. Pursuant to the Copenhagen Accord promulgated at the UNFCCC Conference of the Parties in December 2009, the Secretary-General has established a High-Level Advisory Group on Climate Change Financing.

National Energy Systems

The international community must come together in a common effort to transform the global energy system over the coming decades, and that policy-makers and business leaders must place much greater emphasis on transforming the performance of national (and regional) energy systems over the coming decades. Low-, middle-and high-income countries all face major, albeit different, transformational challenges:

1. *Low-income countries* need to expand access to modern energy services substantially in order to meet the needs of the several billion people who experience severe energy poverty in terms of inadequate and unreliable access to energy services and reliance on traditional biomass. They need to do so in a way that is economically viable, sustainable, affordable and efficient, and that releases the least amount of GHGs.
2. *Middle-income countries* need to tackle energy system development in a way that enables them progressively to decouple growth from energy consumption through improved energy efficiency and reduce energy-related GHG emissions through gradually shifting toward the deployment of low-GHG emission technologies.
3. *High-income countries* face unique challenges. As the large infrastructure investments made in the 1960s and 1970s begin to reach the end of their economic lives, they present opportunities to further decarbonize their energy sectors through new investments in lower-carbon generation capacity. In addition, they will need to reach a new level of performance in terms of energy use.

While different national economies may pursue these transformational paths in distinct ways, there are large potential synergies from international cooperation, joint strategies and the sharing and adaptation of emerging best practices. These include lessons learned from policies and regulations, capacity development, technical standards, best available technologies, financing and implementation approaches, and more coordinated, scaled-up research and development.

By 2030, there is an opportunity for the world to be well on its way to a fundamental transfor mation of its energy system, allowing developing countries to leapfrog current systems in order to achieve access to cleaner, sustainable, affordable and reliable energy services. This change will require major shifts in regulatory regimes in almost every economy; vast incremental infra structure investments (likely to be more than $1 trillion annually); an accelerated development and deployment of multiple new energy technologies; and a fundamental behavioural shift in energy consumption. Major shifts in human and institutional capacity and governance will be required to make this happen. The transformation of energy systems will be uneven and, if poorly handled, has the potential to lead to a widening "energy gap" between advanced and least developed nations, and even to periodic energy security crises. But handled well – through a balanced framework of cooperation and competition – energy system transformation has the potential to be a source of sustainable wealth creation for the world's growing population while reducing the strain on its resources and climate. While there are various possible areas of focus in the broader energy system, AGECC has chosen two specific areas that present immediately actionable opportunities with many co-benefits: energy access and energy efficiency.

Ensuring Universal Energy Access

AGECC calls on the United Nations system and its Member States to commit themselves to two complementary goals:

1. *Ensure universal access to modern energy services by 2030.* The global community should aim to provide access for the 2-3 billion people excluded from modern energy services, to a basic minimum threshold of modern energy services for both consumption and productive uses. Access to these modern energy services must be reliable and affordable, sustainable and, where feasible, from low-GHG-emitting energy sources. The aim of providing universal access should be to create improved conditions for economic take-off, contribute to attaining the MDGs, and enable the poorest of the poor to escape poverty. All countries have a role to play: the high-income countries can contribute by making this goal a development assistance priority and catalyzing financing; the middle-income countries can contribute by sharing relevant expertise, experience and replicable good practices; and the low-income countries can help create the right local institutional,

regulatory and policy environment for investments to be made, including by the private sector.

2. *Reduce global energy intensity by 40 per cent by 2030.* Developed and developing countries alike need to build and strengthen their capacity to implement effective policies, market-based mechanisms, business models, investment tools and regulations with regard to energy use. Achieving this goal will require the international community to harmonize technical standards for key energy-consuming products and equipment, to accelerate the transfer of know-how and good practices, and to catalyze increased private capital flows into investments in energy efficiency. The successful adoption of these measures would reduce global energy intensity by about 2.5 per cent per year – approximately double the historic rate.

Delivering these two goals is key to achieving the Millennium Development Goals, improving the quality and sustainability of macroeconomic growth, and helping to reduce carbon emissions over the next 20 years.

There are also important synergies between these two goals. Modern energy services are more efficient than biomass, and the acceleration of energy access will also contribute to a more rapid reduction in net energy intensity. Increased energy efficiency allows existing and new infrastructure to reach more people by freeing up capital resources to invest in enhanced access to modern energy services. Similarly, energy-efficient appliances and equipment make energy services more affordable for consumers – residential, commercial and industrial. While there is no agreement as yet on the minimum target for universal energy access, the initial steps do not entail significant climate impacts.

For example, IEA's recommended threshold of 100 kWh per person per year, even if delivered through the current fossil fuel-dominated mix of generation technologies, will increase GHG emissions by only around 1.3 per cent above current levels. The impact of this increased energy consumption can be reduced through energy efficiency and a transition to a stronger reliance on cleaner sources of energy, including renewable energy and low-GHG emitting fossil fuel technologies, such as a shift from coal to natural gas. While each goal is worth pursuing independently, there will be clear synergies in pursuing them as part of an integrated strategy.

Although ambitious, these goals are *achievable*, partly because of technology innovations and emerging business models, and partly because of an ongoing shift in international funding priorities towards clean energy and other energy issues. There are also precedents for the widespread provision of both energy access (e.g., in China, Viet Nam and Brazil), and for dramatic improvements in energy efficiency (e.g., in Japan, Denmark, Sweden, California and China) that demonstrate the feasibility of achieving both goals.

Recommended Actions to Achieve the Goals

AGECC recommends the following actions toward achieving the two goals of ensuring universal energy access and of reducing global energy intensity:

1. A global campaign should be launched in support of "Energy for Sustainable Development."

This campaign would be focused on improving access to modern energy services and enhancing energy efficiency, as well as raising awareness about the essential role of clean energy in reaching the MDGs while addressing climate change, promoting economic growth and conserving natural resources and biodiversity. The campaign should ensure that energy is made an integral part of the MDG review process in 2010 as well as other major inter-governmental processes — including those on climate change, biodiversity, desertification, food security, and sustainable development. The campaign should encourage the United Nations and its Member States, other multilateral institutions, and the private and non-profit sectors to take the actions needed to achieve its goals.

2. All countries should prioritize the goals through the adoption of appropriate national strategies.

National strategies should create a predictable, long-term policy environment for investment and a road map for accelerating the establishment of the required human and institutional capacity and delivery mechanisms.

For high-income countries, this may entail: (a) national plans to benefit from the energy efficiency dividend; (b) increased investment in R&D; and (c) more focused commitments to support developing countries in helping to achieve their goals in the areas of both energy access and efficiency.

For middle-income countries, this may involve: (a) national plans to capture the energy efficiency opportunities as an integral part of their

National Appropriate Mitigation Actions (NAMAs) and Low Carbon Growth Plans (LCGPs); (b) targeted interventions to reduce residual pockets of energy poverty; (c) a phased withdrawal of untargeted energy subsidies; and (d) technical support for the energy access and efficiency programmes of low-income countries.

For low-income countries, this may require: (a) national plans to accelerate the deployment and provision of modern energy services; (b) incorporation of these plans, if based on low-GHG emissions technologies, into their NAMAs/LCGPs; (c) re-orienting regulatory policy frameworks, including tariff structures and market regimes, to stimulate business innovation and private sector participation; (d) improvement in the design and careful targeting of energy subsidies; (e) further investment in the capabilities of public utilities; and (f) a phased introduction of low-GHG emitting technologies, as well as energy efficiency measures wherever feasible.

In a broader context, all countries have to work towards: (a) accelerated harmonization of technical standards for energy-using products and equipment; (b) increased R&D investments, especially in technologies that would reduce the cost and GHG intensity of energy services; and (c) trade-related measures that would support market expansion for products that increase energy efficiency or enhance access.

3. Finance, including innovative financial mechanisms and climate finance, should be made available by the international community.

A combination of financial support mechanisms and a significant increase in international finance – both bilateral and multilateral – will be needed to catalyze the existing public sector funding mechanisms and to leverage increased private sector investments, in order to meet the capital requirements needed for providing access to modern energy services and energy efficiency programmes in low- and middle- income countries.

For *universal access to modern energy services* to meet basic needs, it is estimated that $35-40 billion of capital will be required on average per year to achieve basic universal access by 2030.

We estimate that around $15 billion of grants would need to be made available, mainly to cover the capital investment and capacity building required in least developed countries, where national energy investments are likely to focus on overcoming infrastructure backlogs and meeting suppressed demand in productive sectors. In addition, $20-25 billion of loan capital will be required for governments and the private sector above business-as-usual.

For *energy efficiency*, our estimate is that on average $30-35 billion of capital is required for low-income countries and $140-170 billion for middle-income countries annually until 2030 above the IEA's reference case. In general, most energy-efficiency investments are cost-effective. In practice, however, costs of energy-efficiency are typically mostly front-loaded, with the benefits accruing over time, and low-income countries often have access to limited and expensive capital, which they prefer to invest in the cheapest (first-cost) options available to attain their energy goals. This is also a challenge for many consumers – residential, commercial and industrial – who look for investments with quick payback periods of typically 2-3 years. Financial support in terms of innovative financial structuring such as concessional loan finance, loan guarantees and other financial instruments, supplemented by other market mechanisms, helps to address the risks and barriers, and leverages private capital.

To support investment in energy access and efficiency, climate finance could be mobilized through two key strategies:

(a) Funds could be made available from the $30 billion "Fast Start Funding" committed in COP-15 under the Copenhagen Accord for 2010-2012, especially for strategy, policy and capacity development. This could be in line with the Global Environment Facility (GEF), or the newly-established, multi-lateral development bank-administered Climate Investment Funds (CIF) which already has donor commitments of $6 billion. In the medium to long term, the Secretary-General's High-Level Advisory Group on Climate Change Financing could make it a priority to address the financing needs for energy efficiency and low-carbon energy access investments.

(b) In parallel, innovative use of carbon markets could expand the effectiveness of the Clean Development Mechanism and other market-based mechanisms as vehicles for the mobilization of incremental funds.

All support should aim at scaling up financial instruments that mitigate the risk of commercial lending for energy access and energy efficiency, and therefore leverage increased private sector participation over time.

4. Private-sector participation in achieving the goals should be emphasized and encouraged.

In the first instance, this will require the creation of long-term, predictable policy and regulatory frameworks to mobilize private capital. Within this context, major opportunities to enhance private participation may include:

(a) Implementing more public-private partnerships (PPPs) that have the potential to accelerate deployment of technologies that improve energy efficiency and/or enhance energy access (especially on the basis of low emissions). These could be akin to successful PPPs in the global public health arena and could catalyze a scaling up of funding for research, development, and commercial demonstration of low-carbon technologies, especially to close the energy access gap.

(b) The creation of new and innovative investment mechanisms to enable accelerated technology deployment with active private-sector participation – e.g., through a network of regional clean-energy technology centres to hasten the spread of locally appropriate energy technologies.

(c) An expansion of local lending capabilities to scale up investments in energy efficiency and access through local commercial banks and micro-finance institutions.

(d) Many countries have established regulatory and incentive frameworks for attracting private capital into the energy sector. These include a separation of regulatory, generational, transmission, and distribution functions; the announcement of capacity targets; transparent long term tariff offers; and coverage for political risk (but not for economic risk). Successful models could be transferred to other countries through South-South cooperation.

(e) The existing systems could be adapted to the emerging challenges, e.g., by adding special incentives for off-grid areas, the deployment of renewables (feed-in tariffs), and R&D. Incentives for off-grid areas may include the expansion of local lending for energy efficiency and access through local banks and micro-finance institutions referred to under (c) above.

(f) The envisaged technology mechanism under the UNFCCC could also be mobilized in this regard. One approach could be to increase private sector participation in the network of regional clean-energy technology centres to hasten the spread of locally-appropriate energy technologies

5. The United Nations system should make "Energy for Sustainable Development" a major institutional priority.

This may be achieved as follows:

(a) Facilitating energy access and improving energy efficiency should be integrated and main-streamed into all relevant programmes and projects of the United Nations system, and Member States should be encouraged to do the same.

(b) Technical and financial support should be provided to help governments formulate appropriate plans, policies and regulations and develop local institutional capacities to enable their effective delivery, with a focus on "delivering as one" through United Nations country teams, supported and facilitated by UN-Energy.

(c) Existing knowledge networks should be mobilized and new ones built with partners outside the United Nations system to accelerate the transfer of best practices (with respect to modern energy system policies and regulations) by

   i) mobilizing expertise across multilateral, public and private organizations;

   ii) designing targeted, technical interventions;

   iii) providing a registry of donor projects to facilitate improved coordination; and

   iv) creating and sharing diagnostic tools, technical software and know-how for policy-makers and practitioners.

   The UNEP-led Global Network on Energy for Sustainable Development (GNESD) provides a good example of knowledge creation and sharing on energy policy analysis.

(d) A monitoring and evaluation system for "Energy for Sustainable Development" should be created and coordinated to allow dynamic tracking of national (and sub-national, e.g., city) progress over time.

(e) A mechanism for regular global dialogue on "Energy for Sustainable Development" should be established, including a secretariat to manage the process.

(f) A strengthened UN-Energy framework could serve to spur progress toward a number of these objectives.

Energy Access

One of the challenges facing the global development community is that there is no consensus on exactly what energy access means. It is useful to consider incremental levels of energy access and the benefits these can provide. For the sake of simplicity, one can consider three levels of access to energy.

However, we have adopted a broader definition because access to sufficient energy for basic services and productive uses represents the level of energy access needed to improve livelihoods in the poorest countries and drive local economic development. "Affordable" in this context means that the cost to end-users is compatible with their income levels and no higher than the cost of traditional fuels, in other words what they would be able and willing to pay for the increased quality of energy supply.

In practice, achieving universal access to modern energy services by this definition will entail providing affordable access to a combination of energy services that can be classified in three headings:

- Electricity for lighting, communication and other household uses.
- Modern fuels and technologies for cooking and heating.
- Mechanical power for productive use (e.g., irrigation, agricultural processing) could be provided through electricity or modern fuels (e.g., diesel, biofuels).

Importance of Energy Access

Universal access to modern energy services is fundamental to socio-economic development. Without access to modern fuels and electricity it is highly unlikely that any of the objectives of the Millennium Development Goals will be achieved.

A lack of access to modern energy services hampers healthcare, gender equality, education, and poverty alleviation. For example, cooking on open fires and insufficiently ventilated and inefficient stoves that use biomass and coal-based fuels, results in an estimated 1.5 million premature deaths every year, disproportionately affecting women and children. Many times this number of people suffer from debilitating respiratory infections. Women are further burdened by the long distances they need to travel to collect biomass for fuel – in extreme conditions, women in some areas of rural Tanzania walk 5-10km a day collecting and carrying firewood, with loads of over 30kgs. It is also difficult for children relying on inefficient and poor-quality sources of lighting, such as candles and kerosene, to learn after dark.

The ability of poor communities to make productive use of their natural resources, time and human energy is severely hampered by the lack of mechanical power. Low-income households typically spend 7-15 per cent of their income on energy, but in countries where energy sources are more difficult to come by or prices are comparatively high, energy can account for as much as 30 per cent of the household's monthly expenditure. In certain cases, there will be financial benefits from replacing traditional fuels with modern alternatives. Electric lighting in particular offers substantial cost savings over the most common alternatives (batteries, kerosene and candles).

In addition to these development aspirations, universal energy access is also important for the climate agenda. While universal access to basic levels of energy services will have a limited impact on greenhouse gas emissions (IEA estimates suggest that basic universal electricity access would add around 1.3 per cent to total global emissions in 2030), increasing the level of energy provision and consumption for productive uses could substantially increase this. This underscores the importance of the accelerated deployment of low emissions technologies, where possible. This applies to both the supply side (including lower-emissions fossil fuel-based technologies) and the demand side, where energy-efficient end use devices reduce the amount of power consumed. Ensuring access to these technologies and developing new products and services geared to the needs of low-income communities is therefore critical.

In addition, in many cases there are environmental benefits to providing energy access, either through newer, lower-carbon-emitting technologies (e.g., solar LED lighting), or reducing deforestation by replacing charcoal with modern fuels (e.g., an estimated 20 per cent of deforestation in the Democratic Republic of the Congo is driven by demand for fuel wood and charcoal). The acknowledgement in the Copenhagen Accord of the importance of reducing emissions from deforestation and degradation (REDD) may create a link between carbon finance and energy access initiatives that reduce deforestation.

Furthermore, black carbon, a key component of soot from incomplete combustion of fossil fuels and biomass, represents a major part of global

GHG emissions. About 26 per cent of black carbon comes from the residential sector – essentially from incomplete combustion in cooking stoves that burn fossil fuel and biomass. Solar ovens and improved efficiency stoves can achieve significant reductions in black carbon. The potential climate benefits are startling. Eliminating all black carbon emissions from cooking stoves over 20 years would be roughly equivalent to changing every car and light truck on Earth to a zero carbon dioxide emitter.

Universal Energy Access is Ambitious but Achievable

Achieving universal energy access is an *ambitious* goal. The scale of the task is daunting and requires overcoming complex challenges in some of the poorest and most remote locations on the globe. Currently, more than 1.5 billion people have no access to electricity, and up to a billion more have access in name only because their power supply is highly unreliable. An estimated 2.5 to 3 billion people rely on biomass and transitional fuels (coal, kerosene) for cooking and heating.

If recent national trends in energy access continue, over the next 20 years an estimated 400 million people will gain access to electricity. Nonetheless, taking population growth projections into account, the number of people globally without access will stay roughly the same, and in many countries will actually increase. The geographical distribution of energy poverty will shift, with more people (both in absolute terms and proportionally) suffering from a lack of energy access in Sub-Saharan Africa, and a still significant proportion remaining without access in South Asia.

Ensuring universal access to modern energy services will thus involve providing new electricity connections to around 400 million households by 2030, and modern fuels and technologies to 700 to 800 million households over the same period. For electricity, global access rates will need to increase by just over 2 per cent per year, while in Sub-Saharan Africa an increase of 8 per cent per year is needed.

Providing universal energy access will pose a number of critical challenges related to gaps in national and local institutional capacity and governance required to produce, deliver, manage, operate and maintain these solutions (including strengthening the capabilities of public sector utilities to provide improved services for all their customers in a commercially viable manner and without political interference).

Additionally, accessing and allocating sufficient financing will be a major obstacle. In order to stimulate economic growth, many countries will naturally prioritize investment in power sector infrastructure for productive sectors (closing the existing supply gap or improving the existing power sector infrastructure) over providing basic energy access. All around the globe, rural electrification is loss-making, and in the developing world this segment of the population is also often the poorest, with the lowest ability to pay. Subsidies are therefore often required to cover capital and, in some cases, operating costs. If the cost of the minimum energy package to end-users should be no more than a reasonable fraction of their income (say 10-20 per cent), it may be necessary to provide temporary subsides to reach affordability in the short-run before economic development accrues. This provides an additional reason why energy for productive uses is so critical: it increases the ability of end-users to pay for energy services, which is key to the long-term financial viability of such services – a virtuous circle.

At the same time, the goal of universal energy access is *achievable*, if the right elements are put in place. The capital investment required for basic access (roughly $35-40 billion per year to 2030) represents only a small fraction (around 5 per cent) of the total global energy investment expected during this period. While more people need access to modern fuels, the capital costs of closing this gap are substantially lower than for electricity.

It is estimated that, on average, $40 billion annually is required through a mix of financial instruments. We estimate that grant funding of around $10-15 billion a year and loan capital of $20-25 billion a year will be needed, with the remainder being self-financed by developing countries. The incremental investment required to provide sufficient energy for productive uses would be almost entirely for concessional loan capital rather than grant funding. This is because the additional energy capacity will provide people with opportunities for income generation and increase their ability to pay for services, thereby increasing the financial viability of the energy services.

Various sources of international funding and risk tools could be accessed to help finance capital and capacity building costs. These include ODA and other donor funding targeted at the achievement of the Millennium Development Goals; and climate-related finance, which under the Copenhagen Accord is intended to increase to $100 billion a year by 2020 (for both mitigation and adaptation). Existing energy programmes and funds (such as the Renewable Energy and Energy Efficiency Fund (REEF), the

Climate Investment Funds of the World Bank and other Development Banks, and GTZ's Energising Development) can be utilized to administer and distribute finance, but will need to be scaled up significantly. This will require governance structures that better balance the needs of donor countries for accountability and the needs of recipient countries for a stronger voice in how the funding is deployed. There are various successful examples of significant scale in the developing world that demonstrate that the technical, financing and operating challenges associated with expanding energy access can be met, even in the more difficult rural settings. As an example, more new household electricity connections were made in the 1990s than would be required in each of the next two decades to achieve universal access. This extension occurred mainly in Asia (especially China, Viet Nam, and Thailand) but South Africa and Brazil also achieved notable successes in rural electrification.

While the challenge in the future will increasingly be that people who lack access will be more dispersed, more rural, and have lower incomes, and will therefore require targeted subsidies in the face of a limited availability of resources to meet higher capital costs, the technologies and business practices required to overcome these obstacles already exist and are evolving rapidly.

Access to Electricity

As discussed, it is useful to consider incremental levels of energy access and the benefits they can provide when planning electricity access programmes. Typically, electricity usage is initially limited to replacing other sources of fuel for purposes such as lighting, and for other low energy consumption devices such as for charging mobile phones. Other appliances that require more electricity to operate (such as televisions and refrigerators) are typically added as people can afford them.

However, access to sufficient power for productive use is the minimum required to achieve the objectives espoused in the MDGs, as it is this increase in productivity that can improve income generating opportunities. This is in turn key to improving the ability to pay for electricity services, thus improving the financial viability of these services.

Access can be provided either at the community or household level. For example, *community level* access could initially be provided to health clinics, education facilities, and central recharging facilities that can be used for battery-powered devices such as LED lights or cell phones. Importantly, this corresponds to the priorities of many ODA and private donor organizations, as well as the commercial interests of private sector players, for example mobile phone operators.

Similarly, communal productive capacity could be created, for example to provide access to electricity or mechanical power for basic irrigation or for simple cottage industries such as basic manufacturing or agricultural processing.

In other cases, it may be quicker to provide some level of electricity access directly to *households*. These different levels and types of access are not necessarily sequential, and depend on the local context and priorities.

The scale and nature of the access gap and locations involved means that electricity will need to be provided through both centralized and decentralized energy technologies and systems, combining the following three general models.

- *Grid extension.* An extension of the existing transmission and distribution infrastructure to connect communities to power.
- *Mini-grid access.* Linking a local community to a small, central generating capacity, typically located in or close to the community. The power demand points are linked together in a small, low-voltage grid that may also have multiple smaller generating sources.
- *Off-grid access.* Generating capacity provides power for a single point of demand, typically a solar household system (SHS).

Grid Extension

This is often the least-cost option in urban areas and in rural areas with high population densities. If pursued at the regional level, especially in Africa, it also offers the opportunity to tap into significant hydropower potential, providing low-cost clean energy. A number of factors underpin successful grid extension, including strong government commitment, a clearly defined role for national utilities, sufficient central generating capacity to allow for the increase in demand, and a focus on reducing capital costs, *inter alia* by increasing the economies of scale of the connections.

For large-scale grid extension to be feasible, the system needs to be functioning well enough to support the additional capacity and demand and enable recovery of costs. In many developing countries this is not the case and would require a refurbishment of the existing infrastructure (generation and grids), improvement of the performance of the utilities through local capability building, implementing best practices for operational improvements (e.g., loss reduction programmes) and resolving fuel supply issues by ensuring the appropriate fuel supply chains and logistics infrastructure are established. In countries where electricity and primary energy prices are regulated and subsidized, steps would need to be taken towards establishing tariff structures reflective of costs. In addition, in some urban environments issues relating to land tenure and informality would need to be overcome, as authorities are wary of providing access to electricity if this may be viewed as indirectly acknowledging rights to land.

There are a number of compelling examples of successful large-scale grid extension.

- China secured electricity access for almost 700 million people over the second half of the twentieth century to achieve electrification for over 98 per cent of the population by 2000. The plan focused on creating local enterprises. Key factors in China's success were the government's ability to mobilize contributions at the local level and the domestic production of low-cost components.
- Viet Nam achieved extremely rapid electrification, expanding coverage from 3 per cent to 95 per cent of households in 35 years, and increasing connections at a rate of 13 per cent a year. Access to low-cost finance and insistence on cost recovery, through tariffs or from government budgets, were important in achieving its goals.
- In South Africa, excess generating capacity and the good condition of the existing grid formed the basis for Eskom to implement an intensive grid extension programme that achieved electrification of over 2.5 million households in less than seven years.
- In Tunisia, the national government and the national utility committed to making a steady long-term rural electrification effort the national priority for over 30 years.

One environmental challenge is that large-scale grid-based electrification programmes have historically utilized predominantly fossil fuel-based generating technologies. This was certainly the case in China, where electrification was driven by a rapid expansion of coal-fired plants, and in South Africa, where the programme leveraged significant over-capacity that had already been installed. In the medium term, fossil fuels are likely to continue to play a major role. Deploying low-carbon-emitting fossil fuel technology solutions, such as natural gas, carbon capture and storage (CCS), high efficiency coal-fired stations, and exploring even newer technologies will therefore be critical to reduce emissions. Mechanisms both to reduce the costs of some of these technologies and to cover the additional costs often associated with cleaner technologies will need to be developed and implemented.

Mini-grid and Off-grid Solutions

In rural areas and settlements further from the grid, *mini-grid and off-grid solutions* may be more attractive, for a number of reasons. First, they can often be deployed more rapidly than grid solutions. Second, they do not rely on excess generation capacity. Third, there is often a significant potential local business- building and job creation opportunities from these solutions.

The levelized costs of these solutions relative to grid-based solutions depend on a number of factors, in particular the capital cost of the generation technology and distance from the existing grid. Renewable energy technologies, including small hydro, solar, wind and various types of bio-energy, are ideally suited to mini-grid and off-grid applications, especially in remote and dispersed rural areas. While the costs of non-hydro, renewable energy-based sources are typically somewhat higher than fossil fuel-based technologies, the learning curve associated with their increased deployment is resulting in increasing cost-competitiveness.

The key challenges related to both mini-grid and off-grid solutions include significant initial capital investments, the capabilities required to install and maintain these systems, and defining and implementing appropriate pricing systems. These have been successfully overcome in numerous developing countries. For instance, micro-finance and flexible payment options have helped overcome the lack of access to finance, which represents a significant barrier even where communities are able to afford a portion of the capital costs – in Sri Lanka and Bangladesh IDA and GEF set up centrally-coordinated credit systems leveraging existing micro-finance institutions to create flexible payment options for solar household systems (SHS) and village hydro power. UNEP's work in establishing India's Solar Loan Programme (ISLP) and Tunisia's PROSOL illustrates how subsidies, translated into favourable interest rates for loans and administered through local banks and SWH suppliers, can be used effectively to help customers overcome the barrier of high initial capital costs.

With mini-grid systems specifically, there are sometimes addition challenges related to operating complexity and costs, including load balancing. In many cases, however, the value of aggregating supply in a mini-grid at the community level so that it is available for productive use during non-peak hours for household use will outweigh any additional costs. Mini-grids played an important part in Chinese rural electrification, and there are more recent success stories in Sri Lanka and Mali. In order to augment rural electrification, under a GEF funded Strategic Energy Programme for West Africa, renewable energy powered mini-grids linking to productive uses are being established in eight countries.

There are several other important considerations:

- The long-term sustainability of mini-grid and off-grid programmes hinges on developing the capabilities of local participants and ensuring local supply chains are in place.
- Another critical factor is ensuring that rigorous quality standards for equipment and installation are met (e.g., PVs on houses, distribution lines). In both China and Bangladesh, ad hoc inspections were carried out for off-grid technologies to ensure quality.
- Specifically for mini-grids, ensuring that the technology is forward-compatible with later grid connections is important if mini-grids are viewed as an incremental step on the energy access pathway. This succeeded in China, where once-isolated local community grids have become interconnected as the national grid expanded to include them.
- For mini-grid solutions, it is critical that consumers are charged relative to their level of consumption. In Sri Lanka this was overcome by using limiting power boards and charging end-users according to self-selected maximum supply capacities.
- Finally, there is a need for awareness and confidence amongst end-users in the solution itself, and acceptance within the local community before implementation is also essential.

The Right Mix of Solutions

The critical question in electricity access is not which of these solutions should be adopted, but rather in what way a combination of these solutions should be adopted. The optimal choice for each country would be driven by the availability of resources, the regulatory and policy environment, the institutional and technical capacity, and the relative costs of each of these solutions. Each comes with its own set of advantages and challenges, and the highest impact will be achieved when grid, mini-grid and off-grid solutions are appropriately traded off and then combined to resolve the challenges in each different market.

The trade-off between grid solutions, mini-grid solutions and off-grid solutions needs to take into account several critical factors. These elements are not static, however, and decisions about them will need to consider their expected evolution. The following specific issues need to be considered:

- *Level of demand*: The level of energy access required is dependent on the needs of each community as well as contextual constraints, such as climatic conditions. This is also linked to the ability and willingness to pay.
- *Length of time for delivery*: Given the distributed nature of both the mini- and off-grid solutions, and the resulting reduction in other dependencies such as transmission rights-of-way and building new capacity, it will typically be possible to deliver these solutions more rapidly than a grid solution. Rather than relying on the incumbent utility to deliver the grid-based solution, services can be provided by private-sector players. The time benefit is especially relevant when there are shortages in generation capacity, as is the case throughout the developing world.
- *Cost of solutions*: The cost of technologies will differ according to local conditions and available natural resources, and so the least-cost fuel mix and technology options will also vary for any specific community. Different solutions will be cost-optimal for urban and rural communities. In urban areas and peri-urban areas close to an existing grid, the costs of extending the grid are relatively low, while high population densities create aggregated demand. Over time, the non-hydro renewable technologies associated with the off-grids and in particular the mini-grids are likely to have much higher learning curve benefits than the technologies associated with the grids, because they are new technologies. This makes mini- and off-grid solutions even more attractive options for the future.

- *Quality of access provided by technologies*: Grid-based solutions should (in theory) provide 24/7 access. However, depending on the generation base of the mini- or off-grid solutions, they are often unable to provide this access 24 hours a day, as the generation of wind and solar energy depends on weather conditions and battery storage is limited and expensive. Advances in battery storage technology (which are likely to be rapid due to the R&D investment in electric vehicles) will, however, improve this over time. The emergence of more energy-efficient appliances will also make off-grid and mini-grid solutions more acceptable

There could be considerable interim benefits from starting non-electrified households on a low-capacity supply for certain hours of the day as a step towards a longer-term solution. In Peru, for example, the utility offered both solutions, inviting communities to choose between constant grid access in the future and the less-optimal solution providing more intermittent power much sooner. In most situations, consumers opted for more intermittent access earlier.

The private sector could play an important role in providing initial off-grid electricity supply. For example, mobile phone companies currently use diesel generators to provide power for their antennae in rural areas in Sub-Saharan Africa. By installing solar PV systems, mobile phone operators could be able to generate sufficient power for their requirements and excess capacity, which could be used to power the local health clinic or school. This could be utilized as a charging station for mobile phones, thus providing a commercial incentive for the mobile phone company to invest in the additional capacity.

It is also important to recognize the need for flexibility in the strategy for different countries. To this end, strategies should be tailored to each country to maximize local resources and satisfy specific local requirements.

For all types of electricity access, past successes show that no single institutional model reliably provides better success rates than others. Both large-scale vertically integrated utilities and smaller decentralized businesses can deliver the required solutions, using public, private and cooperative approaches, depending on the strength of the existing utilities and local businesses. In all cases, however, a degree of central programme-level coordination is necessary.

Cost recovery is essential for the ongoing sustainability of services. Governments need to decide what tariff structures and cost recovery mechanisms (e.g., lifeline tariffs or cross-subsidies) to put in place based on the ability and willingness to pay, which will vary according to income levels and the availability of alternative energy sources in the different regions. For example, lifeline or free basic electricity allocations are set at 10kWh/month per connection in the Philippines; at 300kWh/month in Zambia; and at 50kWh/month in South Africa.

Access to Modern Fuels and Technologies

There are a wide variety of modern fuels, including natural gas, LPG, diesel and renewables such as biodiesel and bio-ethanol. There are also technology options that are required to make use of modern fuels or use traditional fuels more efficiently, such as improved cook stoves.

The suitability of these options depends on factors such as availability, applicability, acceptability and affordability, including access to finance to cover upfront investments. The declining availability of existing sources of fuel makes switching to modern alternatives a necessity in some places. For example, in many parts of India finding sufficient biomass for cooking is becoming increasingly difficult.

The acceptability of the modern alternative to the end-user is essential, as solutions will only gain traction if they meet users' preferences and needs. In many cases, existing methods meet multiple objectives, so providing a replacement that meets only one of these objectives will prove unacceptable. For example, in the South African rural electrification programme, some communities did not switch to electric cooking stoves even when these were provided for free, as they relied on the coal stoves not just for cooking, but also for heating.

The affordability and people's willingness to pay for modern fuels and technologies largely depends on whether and how much people currently pay for fuel. In many cases, modern fuels cost significantly more than people are currently paying or can afford. Furthermore, significant initial payments (e.g., for improved cook stoves or biogas digesters) and/or the need to buy in bulk (e.g., LPG) present major obstacles to the poor, who do not have access to credit.

Subsidies have been used in some cases to overcome affordability challenges (e.g., LPG programmes in Brazil and Senegal). The challenge with subsidies is that they place a significant strain on government resources, and may be unaffordable to many least developed countries. Furthermore, subsidies often end up providing limited benefit to the people who need them most. They are best used only where necessary, in as targeted a manner as possible (e.g., in Brazil the general LPG subsidy was replaced with discounts as part of a conditional social payment programme – Bolsa Familia).

To illustrate the challenges related to providing access to modern fuels and technologies, we have considered these requirements in the context of cooking needs, focusing on LPG, bio-gas and improved cook stoves. This does not represent the full range of needs or applications for modern fuels; they have been chosen as examples of solutions that have been implemented at scale.

Liquefied Petroleum Gas (LPG)

LPG is widely utilized in cooking applications around the world, providing much more efficient use of energy than traditional biomass. The challenge is that operating costs are relatively high (and subject to global oil price fluctuations). The use of LPG thus becomes a financially viable alternative only where households are already making a significant financial payment for energy (e.g., buying charcoal) or are able and willing to pay for a more efficient alternative. Depending on oil prices and local availability, liquefied natural gas (LNG), dimethyl ether (DME) and ethanol gel could provide viable and cleaner alternatives to LPG. LPG and its alternatives will therefore often be viable in urban areas, where roughly 20 per cent of people without access to modern fuels are located. In addition, the operational delivery of LPG-type solutions in urban areas is typically viable because population densities and available infrastructure make distribution easier than in rural areas.

Still, large-scale LPG programmes in Brazil and Senegal demonstrate that these rural distribution challenges can be overcome, at the same time creating local jobs and livelihoods. Assuming that up to a quarter of the rural population without access to modern fuels could afford to pay for their cooking fuel requirements, LPG (along with its alternatives) may represent a viable option for 25-40 per cent of the global population without access to modern fuels. The sustained impact of this solution is limited, however, by that fact that prices are linked to global oil prices. In Brazil, for example, recent increases in fuel prices resulted in the reversal of the trend of replacing traditional fuels with LPG.

This option should be prioritized where charcoal production is resulting in deforestation and degradation. Even though LPG does produce CO2 emissions, these are dwarfed by the reduction in GHG emissions related to changes in land use, and carbon finance could thus be utilized to cover the additional costs.

Biogas

There is a strong case for biogas where people own sufficient livestock: the dung from two cows typically suffices to meet the cooking requirements of a household. As the fuel is produced on site, there are limited distribution challenges or costs beyond the delivery of the equipment. Even though a higher initial investment is required than for the other options (and access to finance therefore needs to be provided), the absence of ongoing fuel costs mean that the annualized cost over the lifetime of the equipment is significantly lower than that for non-renewable modern fuels (LPG, ethanol gel). Replacing LPG with biogas in Thailand resulted in savings per household of more than $70 per year. This is most relevant in some rural and peri-urban settings, but this solution is more suited to South Asia as livestock in Africa are typically free roaming.

Nonetheless, the market for biogas could feasibly represent a solution for up to 20 per centof the people without modern fuel access. Examples in Nepal and Viet Nam have shown how rapidly this solution can be scaled up. Furthermore, this option reduces greenhouse gas emissions by capturing and burning methane, and carbon finance could therefore be used to cover part or all of the costs. In Nepal, it is estimated that each installation avoids 4.6 tCO_2e/year. At $15/t CO_2e a $250 installation could pay for itself in less than four years.

Improved Cook Stoves

For people who lack access to sufficient livestock and biomass for biogas production and who are unable or unwilling to pay for LPG/natural gas solutions, one further option is to improve the efficiency with which they burn biomass. Here improved cook stoves (ICS) offer a feasible alternative. These stoves provide numerous advantages: they double or triple the thermal efficiency of traditional fuels, reduce the harmful effects of poor ventilation, and may also provide some co-heating. They ameliorate a number of serious health and environmental problems caused by current practices. More efficient stoves are relatively inexpensive ($15-60 per unit/$3-12 per person).

However, experience has shown that higher-quality, more durable models (with associated higher costs) stand a much better chance of sustained impact.

While the success of ICS programmes has often been limited, this appears to be a consequence of poor or ill-conceived business models and inattention to financing realities, rather than any fundamental problem with the concept. For example in Nepal, the limited success was largely ascribed to the fact that there was insufficient promotion, education, monitoring and follow-up. Furthermore, prefabricated models were distributed through a prolonged and difficult transportation process to remote mountainous areas, leading to significant levels of breakage.

For all the modern fuels and technology solutions, increased levels of understanding of their benefits and proper use are essential to ensure uptake. In addition, the development of local capabilities to maintain new technologies (e.g., stoves, biogas digesters) is crucial to success. This should be viewed not as an obstacle but as an opportunity for the creation of sustainable livelihoods. In addition, policy and regulatory frameworks are critical triggers for scaling up investments in renewable energy projects.

Based on the options laid out above and modern fuels projects around the developing world, providing universal access to modern fuels and technologies by 2030 would require an initial investment of $2-3 billion a year. This includes an estimate of both equipment and programme costs.

What is Required for Success?

Based on the lessons learned from programmes around the world to provide access to electricity and modern fuels, a number of building blocks for universal energy access emerge as requirements, at both national and international levels. These will all rely on the mobilization of resources and support at appropriate levels from a range of actors in different countries. In particular:

1. *Policy support from governments*: Governments need to prioritize energy access, set aggressive national targets for universal access, and put in place plans and the enabling environment to deliver them. Successful large-scale electrification programmes are underpinned by government targets and priorities that inform a rigorous planning process. The necessary policies, programmatic capabilities, tariff structures and incentives to support these targets and participation from

the private sector also need to be put in place. These policies will need to be translated rapidly into regulations and legislation. This process should be supported by multilateral organizations, international agencies such as the IEA and IRENA, and non-profit organizations.

2. *Access to financing*: The international community needs to provide financial support to developing countries for meeting the global universal energy access and energy efficiency goals proposed by AGECC. The IEA's reference case estimates that it is possible to provide electricity access sufficient to meet the objectives of the MDGs to the vast majority of the world's energy poor in the next 20 years, for an average capital investment of around $35 billion per year.

   Based on a set of assumptions and using the IEA's reference case for universal energy access provision to provide an understanding of where this funding needs to be sourced from, we estimate that most (55-70 per cent) of the capital costs could be recovered through end-user tariffs (including cross-subsidies, for example as was used in China and South Africa), and could therefore be funded through loan finance. The remainder would need to be funded through international grants (20-30 per cent) and government budgets (10-15 per cent). For example, Kenya charges a 2-5 per cent levy on the national utility's revenues towards a rural electrification fund to subsidize grid extension and SHS projects.

   This equates to concessional finance of $20-25 billion per year to provide loan capital to banks and microfinance institutions to fund capital requirements. Approximately $5-7 billion of this would be passed directly to end-users to enable them to meet the upfront costs of energy access. In addition, $15-18 billion is likely to be required by government, utilities and private developers as loans that could be recovered through future revenues (largely as cross-subsidies). Of this amount, some $3-5 billion per year will need to be made available from the national budgets of lower-middle-income countries that are implementing energy access programmes, while the remaining $10-15 billion per year would need to be provided by international donors. These financing requirements could be partially met from the international climate finance and ODA earmarked for the achievement of the MDGs. It is expected that international finance institutions will have a major role to play in distributing this finance, which will require scaling up existing funding mechanisms, and the development of additional, creative financing mechanisms, like for example the GET FiT programme suggested by Deutsche Bank.

3. *Capacity development*: Resolving the challenges related to access to financing, and reducing the costs of energy access and end use appliances, will not be sufficient to improve energy access without complementary efforts to develop the capabilities and capacities of local institutions for the provision of delivery, quality monitoring, finance, and operations and maintenance services. Such capacity development is needed in both the public and private sectors, and at all levels - national, sub-national and community - and should leverage and build on the expertise and knowledge base that has been developed by multilateral institutions and international agencies.
4. *Utility performance*: Improving the performance of public utilities will be critical for the success of expanding the grid and achieving the universal access target, since utilities in developing countries often have technical losses four or five times higher than their counterparts in developed countries. Expertise from the private sector in the developed and developing world should be leveraged to drive these utility improvements.

Providing global energy access is not a luxury, but a necessity. Lack of access to modern energy services is one of the main factors that constrains development for the poorest populations. Providing access to reliable and affordable energy services is critical for development, and increasing the reliance on clean energy sources for energy access is also important for the climate agenda. Access solutions will vary by geography, by setting and over time. There are many successful examples of access expansion to demonstrate that the ambitious goal of universal energy access by 2030 is achievable.

Energy Efficiency

There is a strong correlation between energy consumption and economic growth, and the term "energy intensity" provides a way of understanding the evolution of this relationship. Energy intensity is the amount of energy used per unit of economic output.

Energy intensity can be reduced in two ways:

- First, higher energy efficiency can reduce the energy consumed to produce the same level of energy services (e.g., a more efficient bulb produces the same light output for less energy input).
- Second, the economic structure of individual markets can shift from high energy intensive activities such as manufacturing to low energy intensive activities and sectors such as services, while maintaining, or even increasing, total GDP.

Since 1990, global energy intensity has decreased at a rate of about 1.3 per cent per year due to both structural effects and physical energy efficiency improvements.

In the future it is clear that a step change in the rate of energy intensity reduction will be required.

Energy efficiency is the key to driving the required incremental reduction in energy intensity. It has come to prominence in recent decades as one of the few "no-regret" policies that can offer a solution across challenges as diverse as climate change, energy security, industrial competitiveness, human welfare and economic development. While it offers no net downside to energy-consuming nations, the opportunities have proved very difficult to capture. In recent decades, however, some developed countries and regions such as Japan, Denmark and California have been able to partially decouple economic growth from energy growth, in part due to major and sustained energy efficiency efforts.

Capturing all cost-effective energy efficiency measures could reduce the growth in global energy consumption to 2030 from the 2,700-3,700 Mtoe forecast to 700-1700 Mtoe. This would represent a reduction in energy consumption growth of some 55 to 75 per cent from the business-as-usual case. It would also have a significant effect in emissions: energy efficiency opportunities make up about a third of the total low-cost opportunities based on currently available technology to reduce GHG emissions globally. (Forestry and agriculture, and a move to low-carbon energy supply, represent the balance of the opportunity.)

In all scenarios, energy demand continues to grow: energy intensity improvements are overshadowed by economic growth. Moreover, an improvement of energy efficiency can also act as an incentive to raise consumption. One reason is that because of energy efficiency improvements, energy services may become cheaper. For example, a more fuel-efficient car may result in more driving. A second reason, especially relevant for developing countries, is that certain forms of energy are supply-constrained. For example in case the latent demand for electricity exceeds the supply, electricity savings because of more efficient equipment can open up the opportunity to use additional electricity-consuming equipment, and the net electricity savings effect is nullified. The combination of the two mechanisms is called the rebound effect. Measurements in developed countries suggest rebound effects in the order of 10-20 per cent of the energy saving, but for developing countries the rebound effects may be more substantial. While the energy savings and carbon saving effect may be partially offset by the rebound, an increase in energy efficiency will result in clear improvements in terms of access, welfare and economic growth.

The vast majority of energy demand growth is expected to come from lower-middle-income countries such as China and India, driven by rapid industrialization and an increasingly wealthy population with a rising demand for cars, household appliances and other energy-consuming products. The energy efficiency savings potential, however, is split almost evenly between high-income countries and the rest of the world, mostly due to the retrofitting opportunities on the large existing stock of infrastructure in the developed world.

In most countries, the untapped potential for improvements is available across both supply and demand. A significant opportunity exists in the power sector in the developing world to improve generation efficiency and reduce transmission and distribution losses, and thereby reduce the amount of primary energy (e.g., coal, gas, oil) consumed for the same output. In many ways, the supply side potential is easier to capture in the short- to medium-term, as there are fewer institutional barriers. Improving power sector efficiency is also directly linked to improving energy access, as discussed above.

On the demand side, there are opportunities across all sectors of the economy to improve energy efficiency by reducing final energy consumption, with the largest opportunities in industry, buildings and transport. For instance, a UNIDO project funded by the Global Environment Facility (GEF) on motor systems energy efficiency in China yielded on average 23 per cent improvement with a payback period of well below two years.

If the full identified low-cost energy efficiency improvement potential were captured by 2030, global energy intensity would decrease by 2.2-2.7

per cent per year. This compares with the IEA reference case of 1.3-1.7 per cent, which is similar or slightly higher than the historic rate. Since this potential is estimated on the basis of currently available technologies, the actual figure could prove to be even larger, taking into account future breakthrough technologies or behavioural change, which could provide substantial additional gains in efficiency.

Based on certain reference case energy efficiency improvement assumptions, in 2030 the remaining opportunity that can be captured in high income countries is spread across industry, buildings and transport, but industry would represent the largest opportunity in the developing world. On the supply side, the power sector mix is projected to change significantly, and substantial efficiency gains will occur due to this change of the mix and the higher efficiency of new plant. To some extent the different energy intensity can be explained through a net export flow of energy intensive commodities from developing to developed countries. In addition, exchange rates play a role; measurements based on purchasing power parity give a different picture than those based on the market exchange rates used here.

To reach the global target of a 2.2-2.7 per cent reduction in energy intensity, developed countries need to reduce their energy intensity by 2.2-2.4 per cent a year on average (almost double the historic rate of 1.2 per cent between 1990 and 2007). Developing countries need to reduce energy intensity by around 4 per cent a year. This is an increase of more than 50 per cent from their historic 2.5 per cent improvement, which is higher than the developed world because of the rapid industrialization and economic growth in some major developing countries. China and India, for example, have had energy intensity improvement rates of 6.4 and 3.6 per cent respectively since 1990. While these numbers cannot be directly extrapolated to the rest of the world, these data do suggest that rapid progress is possible on a large scale.

The type of response towards these goals will differ by sector. In many sectors the nature of the opportunity is similar for both developed and developing countries. For example, there are similar initiatives to improve the efficiency of lighting and appliances, and the fuel efficiency of the vehicle fleet all around the world. In sectors with long-life assets, however, it differs.

In developing countries, much of the energy efficiency potential in buildings, industry and power is associated with greenfield opportunities (i.e., new buildings, new industrial stock). There is a need to move quickly on these infrastructure opportunities: continuing energy-inefficient expansion can lock in infrastructure that will require high energy consumption and carbon emissions for 40 years or more. While retrofit opportunities do exist, they tend to be more expensive. Furthermore, opportunities in the developing world are heavily concentrated in industry, the primary driver of its economic growth.

In developed countries, the energy efficiency opportunities in the near term focus more around retrofitting and upgrading existing infrastructure, or accelerating the retirement of the least efficient assets and replacing them with more efficient ones. Although this is more expensive than capturing the opportunity at the point of construction, it is nonetheless vital if the enormous energy consumption of the developed world is to be tackled. New-build opportunities exist here as well, though this is largely from replacing assets reaching the end of their working life.

Low-income countries represent a relatively small part of the absolute global energy efficiency potential, but still need to get onto the right track and benefit from positive spillover effects from the better-off countries, particularly from standards, learning curves and economies of scale. Critically, improved energy efficiency will assist in providing energy access to more people by making existing capacity stretch further – thereby reducing the total need for investment in new capacity, and keeping energy costs lower.

The analysis and recommendations that follow focus on the developing world, since this is where the UN, the World Bank and donors have a particularly important role to play. It is of course critical that the developed world also takes action to improve energy efficiency. In addition, developed countries can serve as role models and as promoters of energy-efficient measures. They are also likely to shoulder a large part of the cost burden of technology development for energy-efficient products and systems.

Energy efficiency improvement encompasses many different activities across various different sectors:

- Energy efficiency measures in industry include switching away from energy- intensive materials (e.g., clinker substitution in cement), improved maintenance, using efficient burners, and cogenerating power by using waste heat from industrial processes. National policies that set targets and standards have resulted in significantly higher industrial efficiency in Japan and the Netherlands than most other countries. Awareness, training and performance management to change the mindsets of management and staff is also crucial. Special attention should be focused on small and medium-size enterprises and on systems approaches that go beyond the process or technology level.

- The biggest opportunities in building energy savings are improvements to insulation and design (e.g., windows, shell) and efficiency of heating, ventilation and air conditioning (HVAC) systems (e.g. district heating). Denmark and China are examples of countries where significant savings have been achieved through the effective introduction of building codes and standards.
- For short- to medium-life assets such as appliances and lighting, the focus is on switching to more efficient devices such as appliances with low standby power consumption, and CFLs and LEDs rather than incandescent lighting. Making lighting more energy-efficient is often the first efficiency measure undertaken due to the low cost and ease of capture (e.g., as in Bangladesh, Bolivia, China, Cuba, Ethiopia, India, Mexico, Philippines, Rwanda, South Africa, Sri Lanka, Thailand, Uganda, and Viet Nam). Several developing countries have also successfully introduced standards for various appliances, including chillers (Thailand, India, Philippines), electric motors (China), refrigerators (Brazil, Mexico) and air conditioners (Thailand).
- Similarly, in the transport sector, a mix of energy-efficient vehicles provides significant potential, *inter alia* by improving the fuel consumption of the vehicle fleet through improved fuels and engine technology as well as the increased use of all-electric and hybrid electric vehicles. Integrated traffic planning and modern public transportation systems can create significant energy efficiency gains, while concurrently addressing congestion and air pollution. This is especially relevant in rapidly-growing urban areas in developing countries. Bogotá, in Colombia, is a good example. The city created special lanes for buses, introduced a more effective pricing system, and replaced the oldest buses with more efficient models. The project led to a reduced number of buses while maintaining the level of service, and lower fuel consumption per passenger-mile. As a consequence, the project was partly financed by CDM credits.

- The power sector can significantly increase its energy efficiency through implementation of currently available improvements in many forms of power generation, and in improved electric grids that enhance reliability and reduce transmission and distribution line losses. Reducing these line losses requires both improved maintenance and significant capital investment. This is a particularly important opportunity in the developing world, where losses are typically significantly higher.

Benefits of Capturing the Energy Efficiency Opportunity

Much of the recent attention to energy efficiency has its origins in the need to reduce carbon emissions; energy efficiency opportunities make up about a third of the total low cost opportunities to reduce GHG emissions globally. A large number of currently available energy efficiency opportunities are characterized as having "negative cost": in other words, the savings from reduced energy consumption over the lifetime of the investment exceed the initial cost. It is estimated that the total financial savings, or avoided energy cost, of this efficiency opportunity is $250-325 billion a year in 2030.

Additional benefits include the environmental benefit - a reduction of 12-17 per cent of total global GHG emissions in 2030 versus a baseline scenario, which is around a third of the low-cost GHG abatement opportunity - and the economic benefit of reducing the risk of price volatility as a result of demand outstripping supply. When coupled with other low-cost abatement actions such as renewable power and reduced deforestation, this path is compatible with a 450 ppm stabilization scenario.

In addition to the benefits shared by the global community, countries that succeed in increasing energy efficiency can also reap a number of direct benefits at different levels:

- *Governments*. Energy efficiency can ease infrastructure bottlenecks by avoiding or delaying capital-intensive investments in new power supply without affecting economic growth. This is especially important in developing countries, where there are energy supply shortages and significant capital constraints. The IEA estimates savings of $1 trillion in avoided energy infrastructure investment to 2030 if the available energy efficiency potential is captured. Reducing peak load through load management can reduce generation costs. Reducing overall generation through energy efficiency reduces fuel imports (primarily oil and gas), which lowers import dependence, reduces import bills and overall energy costs, and improves the competitiveness of the economy. In sectors with energy subsidies, energy efficiency helps mitigate the burden on the government budget. In terms of project economics, energy efficiency options almost always have positive financial returns and are almost always cheaper than installing new supply.

- *Consumers.* Energy efficiency allows lower energy consumption for the same end-use energy services, which lowers energy costs for consumers - industrial, commercial and residential. This leads to higher affordability, which is particularly important for low-income groups, and creates a more attractive environment for tariff reform. Efficient lighting alone, could save more than $1 a month per household. This would be even more for households that currently rely on kerosene and candles for lighting (the average non-electrified household in South Africa, for example, spends $5-6 per month on lighting). At the same time, reducing energy demand leads to higher system reliability, which in turn lowers outage costs and raises productivity and income.

Energy efficiency can also generate significant employment from additional business activities in the manufacturing and service sectors, such as appliance substitution, public lighting, and other programmes.

Barriers to Capture

However, it is important to balance this view of the benefits against the many barriers and distortions that can lessen the financial gain and make energy efficiency hard to capture. When one moves away from the societal perspective used to calculate the overall energy efficiency opportunity, and adopts a sector or household view, these barriers become clearly evident. The cost of capital, taxes and subsidies all matter in determining the attractiveness of an investment, and transaction costs such as programme and administrative costs can significantly reduce the potential savings on offer. In many countries, energy subsidies distort price signals and present a substantial disincentive to invest in energy efficiency.

There are a number of other major obstacles that apply across both the developed and developing world, though their relative importance varies.

- *Capital constraints* are a particular issue in the developing world. This factor alone impedes, for example, the construction of new power infrastructure that could greatly increase generation efficiency. At a household level too, more efficient appliances are often out of reach due to the higher upfront cost, even if it represents a cost saving over time. This is often compounded by bureaucracy limiting access to financing.

- *A lack of awareness and understanding* of energy efficiency opportunities can limit action on the part of end-users and lead to reticence of lending institutions to fund energy efficiency initiatives.
- *The unavailability of energy efficient technologies* is a major barrier in developing markets and low-income consumers - energy efficient technologies are often targeted only at high-end consumers.
- *The limited appeal of energy efficient technologies*, for example due to poor design or limited features, can deter investment.
- *Agency issues and split incentives* mean that in many cases, the financial benefits of energy efficiency do not accrue to the decision maker. For example, landlords have no incentive to invest in energy-efficient buildings when the benefits go to the tenant, and appliance manufacturers will not adopt efficient technologies unless consumers show themselves willing to pay for them.
- *The lack of capabilities and capacity* in many developing countries to design and implement the required regulations, financing mechanisms and energy efficiency measures is a further obstacle. Even given sufficient capital, many players would currently not be able to capture the full range of available efficiency savings, because they lack the necessary implementation capabilities.

Overcoming Barriers

Barriers can be overcome by a combination of establishing appropriate regulation and standards, setting the right pricing points, easing access to finance, building capability and undertaking informational campaigns:

Policy and regulation. When it comes to practical implementation, experience shows that changing the behaviour of households, businesses and individuals requires a combination of an appropriate regulatory environment and financial incentives. A broad set of policies is required to set standards, reduce transaction costs, align incentives, monitor performance and otherwise overcome market failures.

Codes and standards are a very effective tool. National energy management standards, which have proven successful in OECD countries in delivering significant energy efficiency gains in industry, buildings and transport, can bring worldwide benefits. However, effective tracking and monitoring of the implementation of such standards is critical to success. Energy efficiency codes and standards for lighting and home appliances represent some of the fastest and most easily-realized opportunities, as demonstrated by their being the first resort for countries experiencing an energy crisis, such as Cuba and Ghana. The financial incentive can also be large, as the value of energy saved is high relative to the initial investment cost (more than eight times in some cases). Because light bulbs and appliances typically have a short lifespan, moreover, a focus on standards for new products permeates rapidly through the installed base, capturing most of the opportunity without need for retrofitting.

International action on standards can provide momentum by creating the necessary scale to encourage the private sector to invest in research and development to drive down the costs of more efficient technologies. This will have huge positive spillover effects, benefiting countries that would on their own be unable to drive the initiatives in a cost-effective way. The Lighting Initiative, initiated by the IEA in OECD countries and subsequently by the GEF in non-OECD countries, provides an example of this. As part of this initiative, several countries have banned incandescent light bulbs already, and others have committed to do so over the next two-to-five years. The support of major manufacturers (such as Philips and Osram) has been key to the success of this initiative.

For industry and utilities, energy management standards can play a key role. Many countries have developed their own national energy management standards. These are now being internationally harmonized though the new upcoming ISO 50 001 standard.

Financial incentives, in the form of regulations and tariffs, are often required to catalyze action. Finding the pricing point of energy at which an efficiency initiative will gain traction is critical. Much work has been done by the World Bank and others to find ways to reduce or phase out subsidies without making poor households worse off. This policy alone would have a dramatic impact on energy use. Some countries, particularly in the developed world, apply energy taxes, and these go some way to increasing the attractiveness of energy efficiency investments, often cancelling out the economically inefficient disincentive created by transaction costs or high costs of capital.

Other financial incentives may also be required; including access to concessional finance to help overcome cost barriers or performance-based incentives. Changing financial incentives for utilities – to allow them to earn a competitive rate of return on investments in efficiency – is particularly important, as demonstrated by successful utility demand-side management (DSM) initiatives around the world. Here, regulators mandate utilities to undertake DSM, with energy efficiency costs recovered through utility bills. The benefits of involving utilities in the design and implementation of energy efficiency initiatives include the low capital cost of utilities, facilitating access to capital, and economies of scale from relationships with manufacturers.

Utility DSM has its beginnings in North American regulatory initiatives, but many developing countries – including Argentina, Brazil, India, Mexico, Pakistan, Philippines, South Africa, Sri Lanka, Thailand, Uruguay and Viet Nam – have subsequently implemented DSM programmes in local electric utilities, with associated financial incentives.

Access to finance. Given the substantial capital requirements, a critical factor for success is access to finance. To date, a wide range of financing mechanisms has been used around the world, often in conjunction with multilateral financing through the GEF and carbon mar kets, to enable energy efficient investments. These include credit lines, revolving funds, spe cial purpose funds (including equity, mezzanine), partial credit guarantees and loss reserves, and special purpose vehicles. There are a number of international funds providing financ ing for energy efficiency initiatives, such as the Renewable Energy and Energy Efficiency Fund set up by UNDP and a number of Climate Investment Funds set up by the Word Bank Group.

There are also several examples of successful public-private partnerships for providing access to capital to the end-user, such as partnerships with banks. In Thailand, banks provided the distribution infrastructure for deploying government funds, thereby reducing bureaucracy. The government provided zero interest loans (mainly funded from taxes on petroleum products) to banks, who passed them on as low interest loans to end-users or energy service companies (ESCOs)

Institutional capability and capacity development. Delivering the energy efficiency opportunity will require capabilities to be developed across a variety of public and private sec tor stakeholders, including policy makers, regulators and enforcement officials, utilities, and implementers. Beside the initiatives undertaken by the United Nations agencies and World Bank, such as ESMAP, ASTEA and AFREA, specialized NGOs funded by private donors have also provided critical policy support and capability development for the public sector.

Energy Service Companies (ESCOs) play an important role in providing capacity to implement energy efficiency measures in some countries. These are specialized, for-profit companies designed to overcome a number of the existing barriers to energy efficiency investments while taking on risks related to project performance, and sometimes also credit.

Informational programmes. Education and transparency regarding the benefits of energy saving is also important. This is typically achieved through awareness campaigns targeting the private sector and end-users, followed by more specific measures such as labelling, upon which consumers can base their decisions. The success of Denmark's energy efficiency ini tiatives is attributed in large part to the awareness and public discussion of the topic that resulted from the 1974 oil crisis. In Cuba, campaigns targeting the youth have proven to be extremely effective in raising awareness and changing behaviour around energy use.

The most important insight from the various energy efficiency initiatives mentioned above is that achieving energy efficiency improvements on the scale needed will require an integrated approach, with multilateral organizations, governments, industry and the public sector working in parallel. Implementing one or two of the success factors is insufficient: a broad, coordinated approach where multiple barriers are addressed simultaneously is needed to achieve the "critical mass" needed to shift behaviour. Successful initiatives usually require a combination of policy measures enabled through regulation, standards and incentives, as well as financing, capability building and informational programs.

Chinese energy efficiency building codes highlight the importance of effective management systems and capacity for monitoring and enforcement, in addition to regulation. On the other hand, Ghana's lighting initiative provides salutary lessons on the pitfalls to avoid. Faced with an energy crisis,

Ghana tried to move to low-energy lighting, but managed to capture only half of the target potential. Success was limited by lack of market research to understand the technical requirements (in this case of bayonet versus screw fitting), inadequate training for programme implementation teams, and insufficient awareness of CFL technology on the part of end-users.

What is Required to Capture the Potential?

Achieving energy efficiency improvements on the scale needed will require an integrated approach, with multilateral organizations, governments, industry and the public sector working in parallel:

- At the national level, governments need to set energy efficiency targets, develop strategies to deliver these as part of the energy planning process, and create an environment that enables delivery.
- At the international level, the United Nations should encourage countries to commit to targets and develop strategies to meet them. Energy efficiency requires careful planning and a sustained push if it is to be successful. Unless there is a global call for targets, many governments are likely to deprioritize the energy efficiency agenda and overlook the potential benefits.
- Multilateral institutions (MLIs) need to develop best practice knowledge based on leading national standards and regulation for energy efficiency programmes. Several MLIs and privately-funded NGOs are already providing this type of support, but this could be further improved through coordination and alignment of different countries around single global standards.
- The private sector could be encouraged to place more emphasis on R&D on energy-efficient products in order to improve technology, product concept and economics.
- Utilities can be made major providers and facilitators of customer energy efficiency through regulatory mandates and decoupling efficiency improvements from their income opportunities.
- Increased financial resources need to be made available from both public and private sources to fund the additional capital expenditures required for developing countries to meet their higher energy efficiency target over the next two decades. This capital investment amounts to an average of $250-300 billion a year for developing countries to 2030.

> Assuming that lower-income countries (largely made up of China and India) and upper-middle income countries are able to meet their energy efficiency financing requirements internally, the available funding to meet the financing needs of the low-income countries, where lack of funding is most critical, would need to ramp up from $10-15 billion initially to $45-50 billion per year by 2030. Given the short payback period of many investments (less than five years), loan repayments could quickly be rolled over to fund other projects.

Put simply, energy efficiency can save money and reduce carbon emissions while maintaining economic output. It should therefore be a major global priority. There are roles for multinational institutions, governments, industry and civil society to play in overcoming barriers to action in the short term. Action is needed now so that developing nations are not locked into inefficient infrastructure for a generation by short-sighted decisions taken today.

References

IEA (International Energy Agency). 2000. *World Energy Outlook 2000*. Organization for Economic Cooperation and Development (OECD)/IEA

Marsh, R., Larsen, V. G. & Kragh, M. 2010. Housing and energy in Denmark: past, present and future challenges. *Building Research & Information*, 38(1): 92-106.

Madubansi, M. & Shackleton, C. M. 2006. Changing energy profiles and consumption patterns following electrification in five rural villages, South Africa. *Energy Policy*, 34 (18): 4081-4092

McKinsey Global Institute (MGI). 2007. Curbing Global Energy Demand Growth: The Energy Productivity Opportunity. *MGI publication.*

2

Global Energy Crisis

The world demand for energy is rapidly increasing. We need energy to warm our homes, to cook our meals, to travel and communicate, and to power our factories. The amount of energy available to us determines not only our standard of living, but also how long we live. Detailed statistics from many counties show that in countries where the available energy is 0.15 tons of coal equivalent per person per year the average life expectancy is about forty years, whereas countries in Europe and America where the available energy is a hundred times greater have an average life expectancy of about seventy-five years. It is well to remember that a shortage of energy is a minor inconvenience to us, but for people in poorer countries it is a matter of life and death.

The world energy demand is increasing due to population growth and to rising living standards. World population in doubling about every thirty-five years, though the rate of growth is very different in different countries. The world energy use is doubling every fourteen years and the need is increasing faster still. One of the main energy sources is oil and the rate of production is expected to peak in the next few years. There are still plentiful supplies of coal, the other principal energy source, but it is even more seriously polluting than oil, leading to acid rain and climate change. This combination of increasing need and diminishing supply constitutes the energy crisis. The world urgently needs a clean energy source that is able to meet world energy needs. This is without doubt the most serious problem facing mankind. If we simply let things take their course, the world is heading for a catastrophe during the present century.

Before considering the various energy sources in detail it is useful to list some of the difficulties in doing so. These arise partly from the complicated nature of the subject, which involves a range of scientific and technological specialities, and partly from the fierce political debates that surround it. The only way to assess the various criteria is to express them numerically as far as possible. Without the numbers it is all just a matter of words spiced with emotion, and it is never possible to reach an objective decision. These numbers seldom have the precision of scientific measurements, and some of them are inherently imprecise but it is better to have approximate numbers rather than no numbers at all. It is important to distinguish between precise measurements, reasonable estimates, guesses, commercial or political propaganda, and speculations. The speculations can be plausible and in accord with known scientific laws, or in contradiction to such laws. A further complication is that new scientific data often alter the picture; this is notably the case for climate change. The people providing the information can be completely objective, or they can be strongly influenced by commercial and political concerns.

The criteria used to assess the various energy sources are their capacity, reliability, cost, safety, and effects on the environment. No single source satisfies all these criteria so an energy mix is essential for each country. The optimum energy mix depends on the natural resources of that country, and so there is no general solution; each country must be considered separately.

No energy source is completely safe, so it is relatively easy to make a case against any particular source by emphasising its hazards. What is needed is an objective comparison between the hazards of all the energy sources, based on numbers. This may be done by estimating the numbers of workers killed and injured in the course of producing a stated amount of electricity. This excludes the contribution of long-term effects. It is worth mentioning that the casualties due to energy production are small compared with those due to natural disasters. Thus, for example, the Chinese Seismological Bureau estimated that in the years from 1949 to 1976 about 27 million people died and 76 million were injured following about a hundred earthquakes. Huge numbers were also killed by tsunamis and hurricanes.

Every energy source has to be constructed and maintained, and this requires energy. It is thus some time before the energy produced by a device is sufficient to pay back the energy used initially. This payback time is an important parameter when comparing energy sources, but data are rather sparse.

There is one general classification of energy sources that provides a useful guide. This is the degree of concentration. To do anything useful the energy must be concentrated. Energy sources can be divided into three categories: the concentrated sources wood, oil, coal, gas, and nuclear; the intermediate category of hydro which is partly concentrated by the mountain valleys; and the least concentrated such as wind, solar, geothermal, wave, and tidal. These sources contain vast amounts of energy but it is thinly spread and only becomes useful when it is concentrated.

We tend to think that environmental degradation is a recent problem, beginning only with the Industrial Revolution; but in ancient times, when wood was the main fuel, the forests of the Mediterranean were cut down, often leaving deserts. Later on, many of the forests of northern Europe were also cut down for fuel. Wood together with crop residues and dried animal dung is still the principal fuel for most people in poorer countries. This practice impoverishes the soil and makes more deserts. Other ancient energy sources like wind mills and water wheels, although less polluting, produce limited quantities of power. The windmill is especially unreliable, although the waterwheel has developed into hydro-electric power in modern times.

Before considering the possible energy sources individually it is useful to make a few general remarks. While it is essential to express as much as possible numerically, the limitations on the numbers must be borne in mind. These numbers are vital to a proper assessment but are inevitably approximate. They differ from one county to another, and vary with time as new safety measures are introduced. They include all the direct hazards; in the case of coal for example, they include mining and transport hazards as well as those involved in the day to day running of the power stations. The manufacture of safety devices in factories brings with it more hazards, so it is just not possible to make any energy generating device absolutely safe.

The costs of energy generation vary from one country to another and with the distance from mine to power station, where appropriate. Power stations remain in operation for many decades, and during that time inflation affects the costs. The rate of interest on the dividends paid to the shareholders has a critical effect on the final cost.

The power output from energy generating devices estimated by the designer often differs substantially from what is actually achieved. It is therefore necessary to base figures for the power output and the costs on actual operating experience over a number of years. It is thus practically impossible to evaluate a new device without running it for several years.

It is often said that our energy problems could be solved if we used energy more carefully and avoided waste. There is certainly much that can and should be done. We can insulate our homes to conserve heat and avoid heating rooms that are not used. We can turn the heating down and wear more clothing. We can install energy-saving light bulbs. We can walk or use smaller automobiles and avoid unnecessary journeys.

If everyone were to carry out these and many similar measures the energy use would be much reduced. It has even been suggested that in this way we can reduce energy use by a factor of four. Some of these measures are easy and some are not. It is much easier, for example, to build an energy-saving house than to modify an existing house. Many of these measures, such as insulating our homes, require new materials that have to be made in factories. This inevitably requires energy, and we have to consider how long it will take to recover the energy expended. The main difficulty is to convince people to change their way of life. Certainly we have a serious obligation to do what we can to reduce energy use, but even if we do we will still need to generate large amounts of energy.

Energy-saving measures are most important. They can ameliorate the situation but are not able to avoid the energy crisis. We must therefore consider how the available sources of energy can be enhanced and used wisely.

World Consumption of Coal

Coal, together with oil and natural gas, is one of the fossil fuels, which come from the decay of vegetation many millions of years ago. They are all very reliable sources of energy and are not unreasonably expensive. Their main disadvantage is the pollution they cause, of the land, the sea, and the atmosphere.

The world consumption of coal has risen from 100 millions tons of oil equivalent energy in 1860, to 330 in 1900, to 1300 in 1950 and to 2220 in 2000. In 1950 it was by far the world's largest energy source, but by 2000 it was easily exceeded by oil. The lifetime of world coal supplies is often calculated by dividing the coal reserves by the annual consumption, and this gives about 250 years. However Lomborg has found that this ratio seems to stay the same from year to year, the increased consumption being balanced by the discovery of new reserves.

The main concern about coal is the pollution it causes. A typical coal power station produces as solid waste over a million tons of ash, 21,000 tons of sludge, and half a million tons of gypsum and discharges into the atmosphere eleven million tons of carbon dioxide, 16,000 tons of sulphur dioxide, 29,000 tons of nitrogen oxides, and a thousand tons of dust, plus smaller amounts of aluminium, calcium, iron, potassium, nickel, titanium, and arsenic.

This anthropogenic pollution can be compared with that due to natural causes, such as bush fires due to lightning strikes and volcanic eruptions. Although the short term effects may be severe, the earth has great natural recuperative powers; and once the source of pollution is removed the land, lakes, and seas return to their previous state.

Unlike these natural events the pollution from energy generation builds up continuously, and so the earth cannot recover. The solid waste has to be deposited somewhere, often in the sea, hazarding aquatic life. The atmospheric waste produces acid rain and climate change. The acid rain causes plants and trees to weaken and die, and renders lakes sterile and kills the fish. By the 1980s, nearly 4,000 lakes in Scandinavia were dead and 5,000 had lost most of their fish.

It has been suggested that the carbon dioxide, which is the principal ingredient in atmospheric pollution, could be sequestered, that is put into liquid form and pumped into empty oil wells. This process is expensive and could increase the price of coal by a factor of two or three. Even if this were done, there would still remain the hazards of the other atmospheric discharges.

The vicinity of a coal power station is hardly an object of beauty. The waste from burning the coal is usually stored nearby and forms large, unsightly, and dangerous slag heaps. They are dangerous because after heavy rain they can collapse, overwhelming nearby buildings. This happened some time ago in the Welsh village of Aberfan. The slag flowed over the village school, killing over a hundred children.

Coal is by far the most hazardous of the energy sources. Mining is dirty and dangerous; over 80,000 miners were killed in accidents from 1873 to 1938. A detailed study found that about forty miners are killed to produce a thousand megawatt years of energy, and many hundreds of thousands have had their health permanently impaired by silicosis and other diseases. For all these reasons it is imperative to phase out coal power stations as soon as possible.

World Consumption of Oil

The world consumption of oil increased very rapidly throughout the twentieth century, partly because it is easier to extract from the earth than coal and partly because it is easier to transport by pipeline or tanker from well to power station. It also has a higher calorific content than coal. In 1900 the world oil production was twenty million tons, 470 in 1950, and 3400 in 2000.

The safety of oil occupies an intermediate position with about ten deaths per thousand megawatt-years. This is mainly due to oil well fires. There were 63 accidents in the period 1969-1986, with an average of fifty deaths per accident.

Oil is serious threat to the environment because tankers are sometimes wrecked and the oil discharged into the sea, killing fish and seabirds, and destroying marine plant life. The polluted area soon recovers and it is worth mentioning that more oil pollution is caused by tankers cleaning out their tanks.

The main disadvantage of oil is that world oil production is expected to peak in about ten years and thereafter fall. This may be offset by new discoveries, although no large oilfields have been discovered since 1980. The demand for oil continually increases. Burning oil also produces large quantities of carbon dioxide, just like coal. Furthermore, oil is a valuable chemical with many applications principally as airplane fuels and in the pharmaceutical industry, and so burning it is very wasteful. A further complication is that the bulk of the remaining oil reserves are in the Middle East.

Oil can also be extracted from tar sands. There are enormous deposits in Northern Canada, estimated to be able to yield at least 170 billion barrels compared with about 260 billion barrels in Saudi Arabia. Venezuela also has substantial reserves. The oil is extracted by boiling water, and is an expensive and very polluting process. It costs about $25 to extract a barrel of oil, so the process is economic as long as it remains less than that from oil wells, as is the case at present.

Oil in the form of ethanol can be extracted from sugar cane and from maize. Already there are large plantations growing crops for this purpose; Brazil plans to plant 120 million hectares and an African consortium 380 million hectares. This takes up valuable agricultural land, however, leading to food shortages, and is also highly polluting. It is, therefore, unwise to rely on oil, even from vegetable sources, for our future energy supplies.

Natural Gas

Sometimes associated with oil and sometimes on its own, gas is an attractive energy source. It comes out of the ground easily and can be transported over large distances either by pipeline or less conveniently in liquid form by road, rail, and ship. It is widely used for domestic heating and cooking It is one of the cheapest and safest energy sources, so many gas power stations are now being built. These power stations can be brought into action rapidly and so are useful when dealing with fluctuating demand. Natural gas is also the safest energy source, with an average of half a death per thousand megawatt years.

The contribution of natural gas to world energy consumption has risen from 170 million tons of oil equivalent in 1950 to 2020 million tons in 2000. A large gas field in Siberia now supplies around 20 percent of western European gas. Gas consumption in Britain is rising rapidly and with it the price. The calculated lifetime is about sixty years, but as in the case of coal and oil this is likely to be an underestimate. Ultimately gas production will fall, like that of oil.

Renewable Energy Sources

Recognition of the pollution caused by fossil fuel power stations has led to strong advocacy of what are sometimes termed the "benign renewables." This label is somewhat misleading, as statistics show that they are by no means benign. The word "renewable" implies that they do not rely on sources that are limited in amount; they rely on the practically inexhaustible sun. In all cases the energy available is enormous, but it is thinly spread and therefore costly to concentrate. It is regrettable that this renders most of them uneconomical for large-scale energy generation, except for hydro where nature does the concentrating for us. They have many attractive and valuable features, but the laws of physics are inexorable.

Hydropower

Hydropower (hydro for short) is a well-established and reliable source that supplies most of the electrical power in mountainous countries like Norway and Switzerland. It is however limited worldwide by the number of suitable mountains and cannot ever supply more than about three per cent of the world's energy needs. There are untapped sources in remote areas, but the electricity produced there has to be transported over long distances and the power lines are exposed to attacks by guerillas.

Hydropower is relatively safe, with a death rate of about four per thousand megawatt years. The dams that hold back the water seem so solid that even this hazard is surprising. However, it sometimes happens, especially with earthen dams, that water starts to trickle through small channels, gradually weakening the dam until it collapses. A wall of water then surges down the valley, obliterating everything in its path. If people are living there, a large number could be drowned. In the period 1969-1986 there have been more than eight dam collapses, with an average death toll of more than 200 people. In one case, about 2500 people were killed. The lakes behind the dams provide a habitat for wild life, and they can be popular for boating. However in times of drought the water level falls and exposes ugly bands of mud. In addition, these lakes often inundate picturesque valleys and their villages, and destroy valuable agricultural land.

Wind

Of the remaining renewable energy sources, wind is the most promising. Windmills have been used since ancient times, and now wind turbines are a familiar sight in the countryside. They have several disadvantages, however, the main one being that the wind does not always blow and so the power output fluctuates instead of remaining steady. The fluctuations are magnified because the power output is proportional to the cube of the wind velocity. This means that energy is available only over a limited range of wind velocities; when the velocity is small very little energy is produced, while if it exceeds the safety limit the blades have to be feathered to avoid catastrophic damage.

The total energy in the wind is more than enough to satisfy all our energy needs but this cannot be realized because of the high cost (two or three times that of coal power), the unreliability, and the large amount of land required. It may however make a useful contribution if the costs can be substantially reduced.

Wind power is surprisingly dangerous at five deaths per thousand megawatt-years. This is due to the large number of turbines required, about a thousand, to equal the output of one coal power station. These have to be made in factories by processes which are inevitably hazardous. In addition, there are the hazards of construction and maintenance.

The environmental impact of wind turbines is increasingly recognised. They must be built in exposed positions where they can be seen for miles around. They emit a persistent humming sound which people living nearby find intolerable. Often people who moved to the country for peace and quiet are forced to leave and then find that no one wants to buy their house. Wind farms can also be built offshore but this increases the cost and may pose a danger to shipping.

In spite of intensive work over many years wind power is still uneconomical, and in most cases it relies on massive Government subsidies. It is fair to propose that research continues until this difficulty is overcome, but that until this is achieved it is unwise to deploy wind turbines on a large scale.

It is sometimes argued against wind power that turbine blades kill large numbers of birds, estimated to be about 70,000 a year in the United States. This figure should be put into perspective by comparing it with the numbers killed on motorways, amounting to 57 million per year in the United States, by colliding with glass windows (98 million per year), and by domestic cats (55 million a year in Britain).

At present wind contributes only about 0.2 percent of Britain's energy. The Government has announced that the energy from all the renewables must be raised to 10 percent by 2010. This requires about 8,400 turbines spread over an area of about 1300 square kilometres. There is no hope of doing this, and even if it were achieved there would still be the problem of generating the remaining 90 percent. The situation is very similar in the United States.

Tidal

Some river estuaries are so formed that they experience high tides. When there is a high tide, the sea water flows in, sometimes to a surprising distance from the sea. Around low tide this water flows again back to the sea. If a barrier is put across the river the water flows through pipes to the sea. It is then easy to make this flow rotate a turbine and generate electricity. Such a device has operated in the La Rance estuary in France for many years, producing 65MW. It is reliable, although the peak periods vary according to the moon and not the sun, so the electricity is not always available when it is needed.

A similar though much larger scheme has been proposed for the Severn estuary between England and Wales. It would cost about fifteen billions pounds spread over about ten years to build and would produce about 7GW. The environmental effects are expected to be severe as the whole ecology of the area would be altered. The cost of the energy produced would be about twice that from a conventional power station. It is a practicable but hardly attractive prospect.

Wave

Once again the energy in the waves is enormous, but it is difficult to concentrate. A number of devices to do this have been built, but the output is not cost-effective. One such device, costing over a million pounds, had a power output of 75 kW, enough for 25 domestic electric heaters. Wave machines are, moreover, always at the mercy of storms, which can destroy them in a few minutes.

Solar

The sun pours energy on to the earth at the average rate of about 200 watts per square metre so that the amount of energy that we obtain is proportional to the area of the collectors. It has been estimated that to supply the energy needs of four houses requires a collector the size of a large radio telescope. The sunlight can be used directly to heat domestic water circulating in pipes on the roof. This process is reasonably economic and is widely used. Nevertheless, there has to be an additional source of energy for times when the sun is not shining. On a larger scale it is possible to focus the sun's rays on a boiler at the center of an array of hundreds of mirrors. The steam produced can be used to drive a small turbine to produce electricity. The disadvantage is that the mirrors have to be constantly turned by servomechanisms to keep the sun's rays focussed on the boiler so the whole process is uneconomic.

Electricity can also be obtained using photoelectric cells. These are expensive to make and produce electricity with a low voltage. They are not economic for large-scale generation, but are very useful to generate electricity in situations where the other sources are impossible or impracticable, such as in satellites and traffic signals in remote areas.

Thus solar power has useful but small-scale applications that will certainly be developed further when the cost of photoelectric cells is reduced. It is not a practical economic source of energy for the major needs.

Geothermal

The interior of the earth is hot, and in some places hot water gushes out. This can be used as an energy source, but on a small scale in rather few places. Elsewhere it is possible to drill two nearby shafts, pulverise the rock between their ends, and then pump water down one and extract it by the other. Passing through the rock, the water is heated and is an energy source. However if the shafts are close the heat in the vicinity is soon used up, whereas if they are far apart the water has difficulty in passing from one shaft to the other. Trials show that this process is absolutely uneconomical.

Costs of Energy

In our society, costs are crucial. Even a small difference is enough to ensure the dominance of one product over another. With energy sources the situation is more complex because the choice depends on weighing the advantages and disadvantages of each source. This is difficult because they are often incommensurable: how much, for example, are we prepared to pay for increased safety or to reduce the effects on the environment? Finally, it is impossible to estimate the cost of delayed damage, such as that due to global warming and climate change. These costs could well be the greatest of all.

Causes of Energy Crisis

Market failure is possible when monopoly manipulation of markets occurs. A crisis can develop due to industrial actions like union organized strikes and government embargoes. The cause may be over-consumption, aging infrastructure, choke point disruption or bottlenecks at oil refineries and port facilities that restrict fuel supply. An emergency may emerge during unusually cold winters due to increased consumption of energy.

Large fluctuations and manipulations in future derivatives can have a substantial impact on price. Large investment banks control 80% of oil derivatives as of May 2012, compared to 30% only a decade ago. Kuwaiti Oil Minister Minister Hani Hussein stated that "Under the supply and demand theory, oil prices today are not justified," in an interview with Upstream.

Pipeline failures and other accidents may cause minor interruptions to energy supplies. A crisis could possibly emerge after infrastructure damage from severe weather. Attacks by terrorists or militia on important infrastructure are a possible problem for energy consumers, with a successful strike on a Middle East facility potentially causing global shortages. Political events, for example, when governments change due to regime change, monarchy collapse, military occupation, and coup may disrupt oil and gas production and create shortages.

Historical crises

1. 1970s energy crisis - caused by the peaking of oil production in major industrial nations (Germany, United States, Canada, etc.) and embargoes from other producers
2. 1973 oil crisis - caused by an OPEC oil export embargo by many of the major Arab oil-producing states, in response to Western support of Israel during the Yom Kippur War
3. 1990 oil price shock - caused by the Gulf War
4. The 2000–2001 California electricity crisis - Caused by market manipulation by Enron and failed deregulation; resulted in multiple large-scale power outages
5. Fuel protests in the United Kingdom in 2000 were caused by a rise in the price of crude oil combined with already relatively high taxation on road fuel in the UK.
6. North American natural gas crisis
7. 2004 Argentine energy crisis
8. North Korea has had energy shortages for many years.
9. Zimbabwe has experienced a shortage of energy supplies for many years due to financial mismanagement.

10. Political riots occurring during the 2007 Burmese anti-government protests were sparked by rising energy prices.
11. 2000s energy crisis - Since 2003, a rise in prices caused by continued global increases in petroleum demand coupled with production stagnation, the falling value of the U.S. dollar, and a myriad of other secondary causes.
12. 2008 Central Asia energy crisis, caused by abnormally cold temperatures and low water levels in an area dependent on hydroelectric power. At the same time the South African President was appeasing fears of a prolonged electricity crisis in South Africa."Mbeki in pledge on energy crisis". Financial Times. Retrieved 2008-02-10.
13. In February 2008 the President of Pakistan announced plans to tackle energy shortages that were reaching crisis stage, despite having significant hydrocarbon reserves,. In April 2010, the Pakistani government announced the Pakistan national energy policy, which extended the official weekend and banned neon lights in response to a growing electricity shortage.
14. South African electrical crisis. The South African crisis, led to large price rises for platinum in February 2008 and reduced gold production.
15. China experienced severe energy shortages towards the end of 2005 and again in early 2008. During the latter crisis they suffered severe damage to power networks along with diesel and coal shortages. Supplies of electricity in Guangdong province, the manufacturing hub of China, are predicted to fall short by an estimated 10 GW. In 2011 China was forecast to have a second quarter electrical power deficit of 44.85 - 49.85 GW.
16. The Economist predicted that in the years after 2009 the United Kingdom will suffer an energy crisis due to its commitments to reduce coal-fired power stations, its politicians' unwillingness to set up new nuclear power stations to replace those that will be de-commissioned, and unreliable sources and sources that are running out of oil and gas. It is therefore predicted that the UK may have regular blackouts like South Africa.

Emerging Oil Shortage

"Peak oil" is the period when the maximum rate of global petroleum extraction is reached, after which the rate of production enters terminal decline. It relates to a long term decline in the available supply of petroleum. This, combined with increasing demand, will significantly increase the worldwide prices of petroleum derived products. Most significant will be the availability and price of liquid fuel for transportation.

The US Department of Energy in the Hirsch report indicates that "The problems associated with world oil production peaking will not be temporary, and past "energy crisis" experience will provide relatively little guidance."

Mitigation Efforts

To avoid the serious social and economic implications a global decline in oil production could entail, the 2005 Hirsch report emphasized the need to find alternatives, at least ten to twenty years before the peak, and to phase out the use of petroleum over that time. This was similar to a plan proposed for Sweden that same year. Such mitigation could include energy conservation, fuel substitution, and the use of unconventional oil. Because mitigation can reduce the use of traditional petroleum sources, it can also affect the timing of peak oil and the shape of the Hubbert curve.

Energy policy may be reformed leading to greater energy intensity, for example in Iran with the 2007 Gas Rationing Plan in Iran, Canada and the National Energy Program and in the USA with the Energy and Security Act of 2007. Another mitigation measure is the setup of a cache of secure fuel reserves like the United States Strategic Petroleum Reserve, in case of national emergency. Chinese energy policy includes specific targets within their 5-year plans.

Andrew McKillop has been a proponent of a contract and converge model or capping scheme, to mitigate both emissions of greenhouse gases and a peak oil crisis. The imposition of a carbon tax would have mitigating effects on an oil crisis. The Oil Depletion Protocol has been developed by Richard Heinberg to implement a powerdown during a peak oil crisis. While many sustainable development and energy policy organisations have advocated reforms to energy development from the 1970s, some cater to a specific crisis in energy supply including Energy-Quest and the International Association for Energy Economics. The Oil Depletion Analysis Centre and the Association for the Study of Peak Oil and Gas examine the timing and likely effects of peak oil.

Ecologist William Rees believes that

> "To avoid a serious energy crisis in coming decades, citizens in the industrial countries should actually be urging their governments to come to international agreement on a persistent, orderly, predictable, and steepening series of oil and natural gas price hikes over the next two decades."

Due to a lack of political viability on the issue, government mandated fuel prices hikes are unlikely and the unresolved dilemma of fossil fuel dependence is becoming a wicked problem. A global soft energy path seems improbable, due to the rebound effect. Conclusions that the world is heading towards an unprecedented large and potentially devastating global energy crisis due to a decline in the availability of cheap oil lead to calls for a decreasing dependency on fossil fuel.

Other ideas concentrate on design and development of improved, energy-efficient urban infrastructure in developing nations.Vittorio E. Pareto, Marcos P. Pareto. "The Urban Component of the Energy Crisis". Retrieved 2008-08-13. Government funding for alternative energy is more likely to increase during an energy crisis, so too are incentives for oil exploration. For example funding for research into inertial confinement fusion technology increased during 1970s.

Kirk Sorensen and others have suggested that additional nuclear power plants, particularly liquid fluoride thorium reactors have the energy density to mitigate global warming and replace the energy from peak oil, peak coal and peak gas. The reactors produce electricity and heat so much of the transportation infrastructure should move over to electric vehicles. However, the high process heat of the molten salt reactors could be used to make liquid fuels from any carbon source.

Social and Economic Effects

The macroeconomic implications of a supply shock-induced energy crisis are large, because energy is the resource used to exploit all other resources. When energy markets fail, an energy shortage develops. Electricity consumers may experience intentionally engineered rolling blackouts during periods of insufficient supply or unexpected power outages, regardless of the cause.

Industrialized nations are dependent on oil, and efforts to restrict the supply of oil would have an adverse effect on the economies of oil producers.

For the consumer, the price of natural gas, gasoline (petrol) and diesel for cars and other vehicles rises. An early response from stakeholders is the call for reports, investigations and commissions into the price of fuels. There are also movements towards the development of more sustainable urban infrastructure.

In the market, new technology and energy efficiency measures become desirable for consumers seeking to decrease transport costs.High Oil Prices Boost Energy Efficiency. January 30, 2008 Planet Ark. Examples include:

— In 1980 Briggs & Stratton developed the first gasoline hybrid electric automobile; also are appearing plug-in hybrids.

— the growth of advanced biofuels.

— innovations like the Dahon, a folding bicycle

— modernized and electrifying passenger transport

— — Railway electrification systems and new engines such as the Ganz-Mavag locomotive

— variable compression ratio for vehicles

Other responses include the development of unconventional oil sources such as synthetic fuel from places like the Athabasca Oil Sands, more renewable energy commercialization and use of alternative propulsion. There may be a Relocation trend towards local foods and possibly microgeneration, solar thermal collectors and other green energy sources.

Tourism trends and gas-guzzler ownership varies with fuel costs. Energy shortages can influence public opinion on subjects from nuclear power plants to electric blankets. Building construction techniques—improved insulation, reflective roofs, thermally efficient windows, etc.—change to reduce heating costs.

Crisis Management

An electricity shortage is felt most by those who depend on electricity for heating, cooking, and water supply. In these circumstances, a sustained energy crisis may become a humanitarian crisis.

If an energy shortage is prolonged a crisis management phase is enforced by authorities. Energy audits may be conducted to monitor usage. Various curfews with the intention of increasing energy conservation may be initiated to reduce consumption. To conserve power during the Central

Asia energy crisis, authorities in Tajikistan ordered bars and cafes to operate by candlelight."Crisis Looms as Bitter Cold, Blackouts Hit Tajikistan". NPR. Retrieved 2008-02-10. Warnings issued that peak demand power supply might not be sustained.

In the worst kind of energy crisis energy rationing and fuel rationing may be incurred. Panic buying may beset outlets as awareness of shortages spread. Facilities close down to save on heating oil; and factories cut production and lay off workers. The risk of stagflation increases.

References

Ammann, Daniel. 2009. *The King of Oil: The Secret Lives of Marc Rich.* New York: St. Martin's Press.

Assis, Claudia, 201. Oil futures extend gains amid unrest, *USA Today,* 23 February.

Therramus, Tom. 2010. "Oil Caused Recession, Not Wall Street". Oil-Price.net. Retrieved 21 March 2012.

William Engdahl F. 2012. "Behind Oil Price Rise: Peak Oil or Wall Street Speculation?". *Axis of Logic.* Retrieved 21 March 2012.

3

Energy Politics

Securing energy supplies is vital in a fast-growing world economy. It evokes, for some, a race among countries to control and plunder energy resources - all the faster to burn them in cars, electricity plants, and their own economic growth. The unfortunate consequence is climate change and worsening poverty in some of the poorest parts of the world.

Can energy security, climate security and reductions in world poverty be achieved simultaneously? Different government departments control policies in each area. Energy policy is mostly formulated within the Department of Trade and Industry. Environment policies are mostly fashioned within the Department for Environment, Food and Rural Affairs. Development assistance strategies are planned in the Department for International Development.

Unfortunately, some present policies are heading in this direction. Many governments failed to meet its targets on CO_2 emissions which have been rising not falling for the last four years. The UK government has no coherent strategy for replacing the one third of UK electricity generation which is about to be retired (much of it nuclear). Its equivocation on this is deterring necessary policy commitments and investments in renewables and carbon-neutral technologies. There is no well-functioning single market in gas in the European Union, nor a common European policy towards Russia, yet these are vital to meet the risks emerging as Gazprom purchases downstream energy assets in Europe and Russian policy takes on a geo-political colour.

China and India are key players in all three areas of energy security, climate change, and development assistance, but they have yet to be engaged as serious partners in all the key institutions addressing these issues. The UK, in common with other OECD countries, has failed to prioritize the transfer of low-carbon technologies.

In developing a better set of policies, it is clear that the UK, as a medium-sized power, can do little alone. The UK will need to work closely with the European Union to forge a strategy better to meet its priorities. This means a strategy designed to produce policies for energy, climate change, and development which are mutually sustainable.

The key elements of a better strategy include, first, deeper and more effective European energy markets and policies which are linked to climate change goals. Second, there should be a better European approach to neighbouring energy producers. Third, the EU should build a new compact with India and China, which includes the United States, and works towards all three goals of energy security, climate change, and development. Fourth, UK and EU development assistance policies should be shaped to address climate effects already being felt and the new global politics of energy. The best strategy for reducing the impact of climate change on the world's poor is stringent mitigation. However, with developed countries, including the United Kingdom, failing to mitigate, priority must be accorded to helping developing countries to adapt to climate change. Finally, the UK itself needs a better UK energy policy framework. We elaborate each of these in more detail below.

Energy Security in United Kingdom

Energy security fluctuates as a public concern in the United Kingdom depending upon the appearance of spectres such as wars in the Middle East causing shortages of oil supplies, terrorists destroying gas pipelines, or Russia turning off the tap to Europe's gas supply.

These perceptions reflect the shift away from the UK's comfortable situation of more than self sufficiency in oil and gas, which characterised the 1980s and 1990s. As North Sea production declines, UK self-sufficiency is diminishing. The UK became a net importer of gas in 2004, and, by 2020, imports could make up about 80 – 90 % of total demand. The UK is also expected to become a net importer of oil by 2010. But does this reliance on imported energy supplies translate into an energy security problem?

The UK's dependence on imports does not necessarily raise concerns over energy security – as is clear in the case of coal which constitutes about 15% of the UK's primary energy supply and which is largely imported. Coal does not raise security concerns mostly because it is widely available from reliable sources at competitive prices. Indeed, the UK could mitigate energy security concerns simply by increasing the use of coal, especially in power generation. However, this would increase UK carbon emissions and put UK climate security in jeopardy.

The UK relies not just on coal but on a mixture of oil, gas, and other forms of power generation. In each of these sectors energy security for the UK is essentially about mitigating the risk of adverse outcomes such as supply disruptions or high and volatile prices. That said, the risks are very different depending on the sector under consideration.

In the case of oil, which constitutes about 35% of the UK's energy supply, the main security issues arise at the international level, concerning high and volatile oil prices and broader foreign policy implications which arise from changes occurring in international oil markets.

Oil (including oil from the United Kingdom Continental Shelf) is priced in international markets which are highly developed. Supplies from different sources are easily substitutable. For this reason, even in the face of supply disruptions – e.g. in the Middle East - the main risk for consumer countries is not physical shortages of oil but high and volatile oil prices.

High and volatile oil prices are a serious problem because oil is predominantly used in the transport sector where, short of new technologies, radical reductions in demand for fuel are difficult to achieve. This means that high and volatile oil prices have severe economic effects. For this reason some advocate that the appropriate response to the risk of short term physical disruptions to oil supply is storage, as with the Strategic Petroleum Reserve in the United States.

For security more broadly, the shifting world of global oil politics has major implications for the UK. The rise of new national oil companies and increasing demand for oil from emerging economies such as China and India create new and serious challenges for UK foreign policy and development assistance goals, to which UK energy policy needs to respond.

In sum, energy security concerns and environmental objectives both point towards the need to limit the UK demand for oil. Yet according to the Joint Energy Security of Supply (JESS) working group of the Department of Trade and Industry, UK oil demand will be broadly flat until 2015 when it will begin to rise slowly.

In the case of gas the situation is rather different. As with oil, the UK is heavily dependant on gas – for domestic and commercial heating, for industry and for power generation. Gas accounts for about 38% of total primary energy supply in the UK with about one third of UK electricity being produced from gas. This situation has arisen as gas has substituted other fuels as it has become relatively cheap compared with the alternatives of oil and coal in space heating and coal in generation. A 'dash for gas' occurred in the 1990s as inefficient coal fired generation capacity was replaced with more efficient and cheaper gas fired installations. Since gas produces less carbon emissions than coal and the plant was more efficient, the switch led to a substantial fall in carbon emissions.

The UK's dependence on gas is unlikely to be easily reversed. In space heating and for industrial use a switch back to oil and coal is neither likely nor desirable: the main possibility for energy substitution is electricity. In electricity generation, with high gas prices, a switch to lower priced coal would be beneficial economically and, since coal is in plentiful supply world wide, would be beneficial for security. Such a switch would, however, be completely incompatible with the UK's goals for carbon emission reductions – unless clean coal technologies, and carbon sequestration, can be developed quickly.

The immediate energy security problems associated with gas are different to those of oil – though international concerns over geopolitical risks in producer countries are similar. The main difference is that gas supplies (like electricity) are grid based. Gas is supplied though a network of pipelines (although an international market in Liquified Natural Gas (LNG) is rapidly developing). A consequence is that that gas markets are more localised and heavily dependent on infrastructure.

Switching between suppliers is much more difficult than in the case of oil. This means that the UK's increasing dependence on gas imports raises potentially more serious concerns over the security of supply. Moreover, the gas grid, like the electricity grid, is a natural monopoly – raising issues for competition and regulation.

Gas security is put in jeopardy as a result of both inadequacies of physical infrastructure and imperfectly functioning markets. Hence the recent UK experience of supply shortages and two spikes in gas prices in the winter and spring of 2005/6. UK wholesale gas prices almost quadrupled, costing UK consumers some £1.5 billion. While domestic supply was not disrupted, industrial users did curtail their use and some electricity production was switched from gas to more polluting coal. The problems arose from deficiencies in the UK's infrastructure and particularly from a lack of gas storage. Britain has the lowest level of storage in Europe and is almost completely reliant on a single facility at Rough. One of the factors behind the second spike in gas prices was a fire in the Rough storage facility in February 2006.

The UK is attempting to improve network flexibility and security by linking into the wider European market via interconnectors, such as the pipeline between Bacton in Norfolk and Zeebrugge in Belgium and a new interconnector to the Netherlands which opened in December 2006. Strengthening gas markets poses a trickier problem. The UK has consistently pushed for a more liberal market in energy across the EU as an aid to flexibility and security and tailored its own policies to that aspiration. However, the reality of EU markets is different. To quote the Competition Commissioner Neelie Kroes, there are "serious problems in EU energy markets" and "indicators of real market distortions".

One problem for the current policy of the UK is that continental EU gas is tied up in long-term contracts. There is nothing necessarily anti-competitive about long term contracts. However, it does mean that connection to the wider European market may provide little additional flexibility of supply in coping with demand or supply shocks. This was apparent in the winter of 2005/6 when the inter-connector failed to deliver the expected balancing flow, despite the dramatic rise in prices in the UK. In particular, operating companies were unwilling to release gas from storage early in the winter because of future obligations to their own customers.

The EU Commission is pushing for greater liberalisation but is facing strong resistance from prominent member states. France and Germany have been particularly cool in responding to proposals for pan-European market liberalisation including unbundling the ownership of transmission systems from supply: Michael Glos, Germany's economy minister, said the move would be "very difficult" and might breach the country's constitutional property rights. Francois Loos, the French industry minister, said bluntly: "Our system works." Indeed, Loos has been actively campaigning against liberalisation as undermining energy security vis-a-vis Russia, arguing that the French "national champions" approach is more likely to be effective.

The proposals by the Commission outlined in 'An Energy Policy for Europe' are promising. The European Council supported the proposals in the communiqué issued at their 8-9 March 2007 meeting. However, there is still a long way to go to complete the single European market in energy, and there are still doubts about whether it will happen.

Internationally, there are further risks associated with current UK policies. The most important third party relationship is with Russia – home to over a quarter of the world's known gas reserves and currently provider of nearly 30% of Europe's gas consumption. For the UK, Russia is particularly important via its effects on continental European supplies and prices. With increasing integration, Russian supply policies affect the price of any additional gas the UK might need. Uncertainty about Russia therefore affects energy security in the UK.

There are two principal concerns about Russia. The first is about Russia's apparent wish to use natural resources in the pursuit of political ends. Observers highlight that in recent disputes with Ukraine, Georgia, Belarus, and Lithuania, Russia's commercial and political objectives have been intertwined, and that Russia has sought to influence EU member states through the routing of pipelines, and has applied political pressure for the entry of Gazprom into downstream markets in Europe.

There is also uncertainty about future levels of supply from Russia. There is a widening potential gas deficit in the country as growing demand within the Russian economy is not matched by growth in supply. The main existing fields are probably passing their peak. Gazprom has failed to invest sufficiently, up to now, in new large-scale production. There will be an inevitable delay of several years until the next "super-giant" field can be brought on stream. Meanwhile supplies to Russian power generators have been reduced, and some generating plants have been obliged to switch to more expensive fuels, including fuel oil.

Further reducing Russian supply in the future is the policy of increasing state control over natural resources, which limits flexibility while diminishing and deterring foreign and other private sector investment. Access by independent producers to monopoly-controlled pipelines has been limited, with the result that large volumes of associated gas have been flared – wastefully and with damage to the atmosphere.

Russia has the potential to be a leading actor in the politics of climate change and energy security, and a valuable partner for the European Union. Indeed, until five years ago, the EU hoped that it was developing with Russia "a genuine strategic partnership, founded on common interests and shared values" (EU/Russia Partnership and Cooperation Agreement, 1997). The G7 decided in 2002 to admit Russia to full membership of G8. However, since 2003 hopes for a strategic partnership have faded as Russia's leaders have made clear that they do not want a partnership on the basis currently offered, and previously envisaged in the 1997 Agreement. For the time being, the EU will not be able to rely on Russia as a partner, to assume that commercial issues will be handled under law-based free market principles, or fully to realise the manifest potential for synergies between the European and Russian economies.

Existing UK policies are attempting to overcome these risks in two ways. At the national level, energy security is being improved by the construction of infrastructure, including increased storage, interconnection to the continental gas network, the new Langeled pipeline from Norway and the construction of LNG terminals. At the international level there has been some diversification of gas suppliers with new supply deals with Norway, Qatar and the Netherlands among others. These measures have undoubtedly improved Britain's access to gas and the diversification of potential sources of supply should improve security. However, at both the national and the international level, risks remain.

Many issues raised about gas arise in the case of power generation by other means: such as the reliability of the infrastructure and the adequacy of generating capacity. However, so far at least, security issues have been less to the fore. Liberalisation of the European market in electricity generation is more advanced than in the case of gas and perhaps more likely to succeed.

A key issue concerns the mix of fuels for future generating capacity, especially as up to one third of present generating capacity is due for replacement in the next five to ten years. This represents a major opportunity for moving to less carbon-intensive generating capacity.

The future mix of fuels is also the most important factor affecting the likely demand for gas in the UK. Uncertainty over the framework of energy policy, as well as equivocation over nuclear power, risks hindering investment, or, at the least, could lead to sub-optimal responses. A clear and credible framework is necessary if the UK's objectives for energy security and for the abatement of carbon emissions are to be reconciled.

Accelerating Climate Change

Over the past five years, albeit belatedly, tackling climate change has steadily become a higher and higher priority for the UK and many other countries' governments and international organisations. The G8 Gleneagles Declaration stated that – "Climate change is a serious and long-term challenge that has the potential to affect every part of the globe ...the world's developed economies have a responsibility to act."

Actions are required to reduce high and rising concentrations of emissions of CO_2, methane and nitrous oxide in the atmosphere. As the Gleneagles Declaration admits, the UK and richer countries have a particular responsibility to take action. The CO_2 stocks in the Earth's atmosphere originated overwhelmingly in the rich world and rich countries still account for the bulk of emissions. Although developing countries are 'catching up' on an aggregate level, they are much further back on a per capita basis. Finally, richer countries are best placed and best resourced to make a difference including through support for technology and changes in policies in developing countries, who will otherwise in the future be the largest emitters .

For Britain, climate change threatens direct damage through higher sea levels, more extreme weather and disruptions to ocean currents. This is recognized but inadequately addressed in the UK Climate Impacts Programme (UK CIP).

Indirectly Britain will suffer from the overseas effects of climate change, such as increased pressure for migration and increased conflicts over scarce resources. Far more devastating still will be the effects of climate change – some already being felt – on the UK's development partners.

The UK government has a publicly stated target to reduce UK emissions by 20% by 2010 and 60% by 2050. The UK is also committed to seeking a new international agreement on a framework for reductions to ensure cooperation beyond Kyoto, and has been an active supporter of the

EU efforts to reduce carbon emissions. In 2005 the government commissioned a report on the economics of climate change by a team headed by Sir Nicholas Stern. The result was a widely publicized statement of the risks posed by climate change and the rationale for taking action now to mitigate it.

Strikingly, the UK is failing to meet its own targets. Indeed, even the government concedes that it is now unlikely to meet the target of a 20% cut of 1990 emissions levels by 2010. National performance on climate change has been disappointing, and is out of step with the UK's international profile as a leader in the field. Britain's CO_2 emissions have been rising, not falling, for the last four years, in spite of the commitment to combat climate change. Britain will probably hit its targets under Kyoto for greenhouse gas emission reductions by 2012 , but this owes more to de-industrialisation in the 1990s, the closure of most of the coal industry, and the market-led and commercially-motivated shift to gas power generation, than it does to government policy.

European policies give no greater cause for confidence. The EU Emissions Trading System (ETS), which was designed to ensure the EU met its Kyoto reductions through capping the overall level of carbon output, has proved ineffective in its first phase, admittedly a trial period of three years. In phase 1 too many permits for carbon emissions were issued making the 'capping' largely meaningless and causing the price of carbon to collapse in April 2006. European Commission commitments to subject national action plans (NAPs) to more scrutiny in the second phase are welcome, but represent a major challenge in facing up to powerful national interests. Furthermore, the EU ETS is a limited scheme, which is focused on industrial producers of carbon and does not address the 50% of emissions produced by activities such as domestic heating and transport. The fact that the scheme expires in 2012 weakens its potential for sending long-term signals on CO_2.

Finally, the international regime is also making very slow progress. Politically, the Kyoto Protocol was important as it marked the first step towards a multilateral response to the problem of climate change. However, the level of ambition for cutting CO_2 was set too low, and two OECD countries (the US and Australia) and the major developing countries are outside the scheme. The Kyoto framework has been a relatively ineffective mechanism, and prospects for its extension are so far not encouraging. The United Nations Framework Convention on Climate Change (UNFCCC) meeting in Nairobi in November 2006 could only agree that there needs to be a new agreement before the present Kyoto framework expires in 2012. The failure to agree a formal timetable for negotiation can only erode hopes of serious international action through this route, casting some doubt on the efficacy of universal multilateral solutions.

Looking to the future, it is clear that international efforts to limit climate change will depend heavily on what happens in China and India. As the fastest growing major economies in the world, China and India also have the fastest growing demand for energy resources, as they seek to sustain levels of economic growth needed to tackle poverty. Their struggle to secure their own future energy sources through long-term supply deals will affect international markets (for oil in particular). Faced with the need to power their booming economies, they are choosing coal over the alternatives because of its relative price advantage and the lack of feasible alternatives which meet their needs in the short-term. But this carries serious potential consequences for climate change. In spite of all this, a serious engagement with China and India, beyond or outside the negotiations on climate change at the UNFCCC, is lacking.

Exacerbating Global Poverty

The dramatic effects of climate change on developing countries – and in particular Africa, low-lying countries such as Bangladesh and small island states – is not yet fully understood. In some regions – notably southern Africa and eastern Africa – models predict a combination of rising temperature and lower rainfall. This has disastrous implications for crop productivity, with declines in excess of 20% predicted for maize in countries such as Mozambique and Malawi. Countries with large river-basin populations – including Bangladesh, Vietnam, and Egypt – face the prospect of large-scale flooding and displacement. A rise in sea levels of 1 metre could displace up to 250m people. Meanwhile, increasingly violent storms and floods will affect a large group of countries, raising the spectre of more catastrophic events such as the 2005 Mumbai floods and the 2005 hurricane season which devastated countries such as Haiti and Nicaragua (as well as New Orleans).

Climate change outcomes will interact with wider ecological pressures. Many of these pressures will be transmitted through water systems. Warming at around 2.5°C has been estimated to place an additional 1.6bn to 1.9bn people at risk of water stress by 2055. Glacial melt, reduced rainfall in tropical areas, and rising temperature will add to water stress in many parts of the world. In the short-run, glacial melt will expose people in countries like Nepal to increased risk of flooding. In the long-run, it will reduce the flow of rivers that feed vital irrigation systems in northern India and western China, jeopardizing hundreds of millions of livelihoods. In the Andes, retreating glaciers threaten to cut off water supplies to entire cities such as Lima.

The UK has one of the most ambitious development assistance programmes in the world. It has doubled aid commitments in 10 years, to £7bn per year by 2007, and led international efforts to reach the Millennium Development Goals. Those efforts are already being jeopardized by climate change. Yet (to quote a recent Parliamentary Report) UK development assistance policies lack a coherent approach to climate change. Similarly, while other bilateral agencies are increasing their aid, most are failing to appraise the impact of climate change on their projects, let alone on wider human development goals. The floods that devastated Mozambique in 2006 and 2007 provide a timely reminder of the risks. Apart from devastating the lives of thousands of people, these floods destroyed a large number of schools and health centres part-funded by UK aid.

We know, from the recent experiences of Kenya, Malawi and Ethiopia, that droughts not only reduce income, with attendant implications for nutrition, access to health and education; they also wipe out the assets that provide security against future risks. These outcomes set in train long-term cycles of disadvantage that hold back prospects for accelerated human development and progress beyond the MDG target date of 2015. Working to achieve poverty reduction and human development in this environment will be like climbing a downward escalator.

The UK Department for International Development has started to bring climate change adaptation to the centre of the development aid agenda. It has supported work by the Hadley Centre to improve climate monitoring in sub-Saharan Africa – and it has linked up with other agencies to develop the analytical tools required to identify programme level responses. That said, both Britain and the wider multilateral system need more actively to address adaptation problems. The current multilateral aid effort is channelled through the Global Environmental Facility. This suffers both from chronic under-financing (less than $30m annually) and a cumbersome project-based approach.

Alongside the effects of climate change, developing countries are also being hit by high energy prices and the new scramble for resources. For energy-importing countries, higher prices are wiping out recent increases in aid. Poor countries tend to dedicate a higher proportion of their GDP to energy imports (e.g. sub-Saharan African countries spent 14% of their GDP on fuel imports in 2000), to use energy less efficiently, and to be less able to switch to alternatives. As a result, the IEA estimates a $10/barrel increase in the price of oil costs Sub-Saharan Africa more than 3% of its GDP. And the price has risen by $40 per barrel in the last four years.

Even for energy-exporting developing countries high energy prices bring risks. In Africa, some fourteen countries are currently enjoying windfalls that are massive relative to any foreseeable aid inflows but the past experience of such revenue windfalls, though highly variable, is on average unfortunate: revenues are often not harnessed for sustained development. Instead, after a short period of time, they tend to increase corruption and even to slow growth.

In recent years, the UK has taken a lead in strengthening governance and pro-poor development in aid-receiving countries. However, the challenges of energy security and climate change may wrong-foot UK development assistance, cancelling out the positive effects of new aid. Pro-poor strategies may fail if stunted by high oil prices and energy poverty. Good governance may collapse in the scramble for energy resources. Development gains may be wiped out by climate change.

Pan-European Energy Policy

Until now, UK support for a pan-European energy policy has been equivocal, and most of Britain's actions have been focused on the domestic infrastructure. The Task Force believes that the UK would better meet its own energy security goals (without damaging its other policy goals) by working towards a more coherent European energy framework. Furthermore, energy security and climate stability are goals which can only be achieved through acting at the regional and at a global level, and Britain can do that most effectively through the EU.

Britain should take the opportunity provided by Commission President Barroso's initiative to develop a new European energy policy. However, it must be careful not to encourage the kind of energy policy which the EU has proposed in the past, one which emphasizes centralised control. Instead, the policy should be one which institutionalises co-operation on energy policy and focuses on networks, infrastructure and on the active application to the energy sector of existing EU competition rules.

The expiry of the current EU Emissions Trading Scheme in 2012 creates risks and opportunities. The risk is that the EU members states will fail to agree to a longer-term, more robust regime. The opportunity is to bind ETS quotas directly to the quantitative reduction targets agreed among EU members.

We would highlight the following priorities for the UK, working with the EU, to reduce the risks we have identified.

First, the completion of the physical European grid for gas and electricity.

Second, common carrier obligations (where owners of transit infrastructure are required to provide it for use equally to all suppliers) need to be established for all EU states, ensuring that individual European states do not seek to guarantee energy security for themselves at the expense of others.

Third, the single market for energy should be completed, including unbundling networks and enforcing competition law and control of state aids against "national champions".

A fourth need is for effective EU arrangements for sharing storage in times of crisis, and for other mutual assistance agreements.

Fifth, a major EU initiative should be launched to provide centralised support for energy research and development focused on viable technologies, especially renewables and carbon capture and sequestration, for which public funding for demonstration projects is a matter of priority.

Finally, the EU should expand its role in standard-setting, building on recent achievements such as the agreement on car emissions by extending the same approach to fields such as energy efficiency.

Crucially, if the emissions reductions targets proclaimed by the European Council are to be met, the EU's carbon trading scheme must be reformed. Quotas must be realigned with the "20% by 2020" target, and set at levels sufficiently demanding to ensure that the target is met. Under the initial scheme, carbon capture and sequestration earns no credits; this defect should be corrected, to encourage investment in CCS technology (e.g. linked to power generation).

Energy Producing Neighbours

Changes in EU external relations with both Russia and North Africa are key to improving the energy security of the continent. The EU's handling of Russia over the past few years has been muddled and incoherent. This is not least because although Russia provides around 40% of overall EU gas imports, different parts of Europe are more directly reliant than others – Britain for example gets just 2% of its gas from Russia; Southern Europe is more reliant on North African suppliers; but Eastern European members get more than 75% of theirs from Russia.

The EU needs a much more clear-sighted and united approach. Completion of an EU single market in energy and physical networks, as proposed above, would make this easier.

Efforts to persuade the Russian Government to accept the Transit Protocol to the European Energy charter appear to have failed. A new basis for negotiation is needed. Given Gazprom's dependence on profits from exports to the EU, Russia has a strong incentive to continue supplies. EU Governments and companies should be encouraged to cooperate in the development of Russia's energy potential, but not on terms which allow Russia to manipulate prices or dictate to the market. There are serious risks that if companies such as Gazprom are permitted to purchase downstream energy assets in Europe before an EU-wide single market has been created, they will subsequently use their position to manipulate prices.

Diversification of energy supply is another key part of a sound European strategy. Pipelines from North Africa and Liquid Natural Gas (LNG) imports from further afield provide the most likely options. Algeria is already a major source of gas imports for the EU, and new findings in North Africa are promising and should be capitalised upon. It is important to note that LNG is an energy source that potentially is more flexible. The development of partnerships with suppliers of LNG, not least in North Africa, is thus important.

Energy relations with North Africa should be further developed using the Barcelona Process or Euro-Mediterranean Partnership (which promotes trade and political co-operation between the EU Member States and the Maghreb and Mashreq countries on the southern littoral of the Mediterranean, and already has a focus on energy). Alongside trade and gas supply deals, EU development assistance for energy projects in the

Mediterranean region should be enhanced. Europe's development programmes for the Mediterranean (principally MEDA) are already being used to support energy initiatives, though at a low level of financial commitment. More significantly, the European Investment Bank's loan facility for the Mediterranean region has been used to considerable effect principally for energy infrastructure. So far 15 loans have been advanced (including one to Turkey) for a total amount of almost 2 billion euros. This financial activity should be extended especially to cover clean technologies and natural gas supply. Indeed this should be part of the EU initiative for fast-tracking the use of carbon neutral technologies referred to above.

The European Investment Bank (EIB) could play a more significant role in the future in increasing gas supplies for example, from Egypt, (both LNG and linking the so called Arab gas pipeline that joins that country, Jordan, Syria and Turkey through Greece and the Balkans to the rest of Europe) and from Algeria and Libya, both of which countries are aiming to double supply. In the first place supply could be via offshore direct pipelines to Spain and Italy and subsequently directly to Italy. Many of the further significant developments elsewhere in the region will depend on politics and security, but in time there is great potential for supplying gas from both Iraq and Iran through Turkey.

A New Compact

The energy needs of China and India are large, and however much they expand their own generation capacity by renewables and nuclear, most of their energy demand will have to be met by conventional sources. In electricity generation their dependence on coal increases the risks to climate stability. In 2004 the International Energy Agency (IEA) estimated that 1400 1GW coal-fired power stations would be built before 2030, with 600 of these being built in China. The climate effect of such a programme (using current technology) would be disastrous. The dependence of these economies on oil for transportation is also fast-growing with additional consequences of international oil markets.

China is increasingly concerned – for its own national environmental and health reasons – to clean up its energy sector. The US and the EU have clear interests in an energy partnership with India and China, building on recent APEC energy activities and in US energy technology initiatives on hydrogen and fuel cells, carbon capture and storage, nuclear, and bio-fuels.

As yet neither China nor India are members of the International Energy Agency. Their engagement in this and other international forums dealing with these issues strikes us as a vital first step. Beyond this we propose a "new compact" between the EU, US, China and India to drive lower-carbon energy use in all these countries, and to support mutual energy security. Such a compact would recognise that climate change and energy security are shared dilemmas which both the developed and developing worlds have an interest in resolving.

The new compact should encompass joint development of energy technology, focusing on clean coal, carbon capture and sequestration, and renewables. These technologies are still new and are not sufficiently invested in by public or private sources, given their potential for contributing to both energy security and climate stability. There should also be agreement on common technology standards for energy production e.g. making clean coal and perhaps even carbon-neutral coal. Many of the technologies have been successful in demonstration projects. The challenge is to reduce the timeframe for facilitating their entry to market.

Enabling China and India to 'leapfrog' to more advanced clean coal technologies is critical. Potentially, this is scenario with benefits for everyone. While the capital costs of super-critical coal power plants are higher, the efficiency gains are considerable. Users in China and India would benefit from these gains – and the world would benefit from reduced carbon emissions. Given the global nature of the benefits, there is a powerful case for a multilateral financing mechanism to facilitate technological upgrading – and Britain should play a leading role in brokering such an agreement. Alongside technological investments and standards, financial transfers will be required to support implementation of technologies which are less efficient and more expensive than existing alternatives.

We see the US as an important partner, relations with whom need to be handled very carefully. Dramatic changes are occurring in US energy policy. Energy security has quickly become a high priority issue attracting bipartisan attention. The Bush administration's initial policy of denial on climate change is being eroded. Although until now the Bush administration has largely de-coupled climate change and energy security issues, this may also be changing.

Energy security has been the heart of the President's last two State of the Union addresses. In 2006 the President spoke of the country's "addiction to oil" and in 2007 of the need to reduce gasoline usage by 20% over the next 10 years. The planned reduction will come from raising vehicle emissions standards for the first time since 1992, and by setting aggressive national targets for the use of alternative fuels and especially ethanol. The 2007 address also included plans to double the strategic oil reserve to 1.5bn barrels by 2027.

While US policy-makers are beginning to focus more on climate change, energy security solutions are threatening to dominate at the expense of climate change concerns. US policy on biofuels and ethanol is a case in point. The US is committed to producing as much as possible to reduce reliance on fuel imports with the President having recently set a target of 35 billion gallons per year by 2017. The consequences for climate change are not good. Using current corn-based technologies, to grow 1.3 units of energy from ethanol takes 1 unit of fossil fuels. A more economic option would be to import ethanol from Brazil, where it is produced from sugarcane, but heavy tariffs have been imposed to prevent this and will stay in place until 2009 in spite of the new US-Brazil strategic partnership on biofuels concluded in March 2007. A more far-sighted option would be to encourage research in second generation, cellulosic biofuels.

US policy on ethanol and biofuels may also have adverse consequences for development goals, driving up global corn prices, and adversely affecting food-importing developing countries. Use of cellulose would, by contrast, support, rather than conflict with, farming for food.

On climate change, movement has been slower than on energy security, but perceptible change is occurring at the centre. The 2007 State of the Union address was the first time President Bush has used the term "climate change", although he is unwilling to engage in setting targets for reducing greenhouse gas emissions, or international efforts so to do. The new Democratic Congress are putting more pressure on the Executive branch to act, and across the US, a growing coalition is calling for more specific action on climate change.

Individual states and cities have taken measures to address climate change. California has been among the leaders, in August 2006 voting for a state-wide cap and trade system for greenhouse gas emissions from major industries and targeting a 25% reduction by 2020. California also led on raising vehicle emissions standards and mandating the use of renewable energy, moves which have been replicated across the nation. Action by states has also galvanised business leaders, who are concerned about a complex web of state regulations emerging due to the lack of federal leadership.

Building on the policy shifts occurring in the United States, it will be imperative to include the US in international efforts to secure both energy and the global climate. A deal which included the US, EU, India and China would not exclude the rest of the world, but would instead focus first on where the problems are greatest.

References

Carlsson, B.; R. Stankiewicz. 1991. "On the nature, function, and composition of technological systems". *Journal of Evolutionary Economics* 1: 93–118.

Freeman, C. and F. Louca. 2002. *As Time Goes by: From the Industrial Revolutions to the Information Revolution*. Oxford: Oxford University Press.

International Energy Agency (IEA). 2009. *Key World Energy Statistics 2009*. Printed in Paris, France by Soregraph.

Jacobsson, S.; A. Bergek. 2004. "Transforming the energy sector: the evolution of technological systems in renewable energy technology". *Industrial and Corporate Change* 13 (5): 815–849.

Worthington, R. K. 1984. "Renewable Energy Policy and Politics: The case of the windfall profits tax". *Policy Studies Journal*: 365–375.

4

Renewable Energy Policies

Renewable energy is energy which comes from natural resources such as sunlight, wind, rain, tides, waves and geothermal heat, which are renewable (naturally replenished). About 16% of global final energy consumption comes from renewables, with 10% coming from traditional biomass, which is mainly used for heating, and 3.4% from hydroelectricity. New renewables (small hydro, modern biomass, wind, solar, geothermal, and biofuels) accounted for another 3% and are growing very rapidly. The share of renewables in electricity generation is around 19%, with 16% of global electricity coming from hydroelectricity and 3% from new renewables.

The use of wind power is increasing at an annual rate of 20%, with a worldwide installed capacity of 238,000 megawatts (MW) at the end of 2011, and is widely used in Europe, Asia, and the United States. Since 2004, photovoltaics passed wind as the fastest growing energy source, and since 2007 has more than doubled every two years. At the end of 2011 the photovoltaic (PV) capacity worldwide was 67,000 MW, and PV power stations are popular in Germany and Italy. Solar thermal power stations operate in the USA and Spain, and the largest of these is the 354 MW SEGS power plant in the Mojave Desert. The world's largest geothermal power installation is the Geysers in California, with a rated capacity of 750 MW. Brazil has one of the largest renewable energy programs in the world, involving production of ethanol fuel from sugarcane, and ethanol now provides 18% of the country's automotive fuel. Ethanol fuel is also widely available in the USA.

While many renewable energy projects are large-scale, renewable technologies are also suited to rural and remote areas, where energy is often crucial in human development. As of 2011, small solar PV systems provide electricity to a few million households, and micro-hydro configured into mini-grids serves many more. Over 44 million households use biogas made in household-scale digesters for lighting and/or cooking, and more than 166 million households rely on a new generation of more-efficient biomass cookstoves. United Nations' Secretary-General Ban Ki-moon has said that renewable energy has the ability to lift the poorest nations to new levels of prosperity. Carbon neutral and negative fuels can store and transport renewable energy through existing natural gas pipelines and be used with existing transportation infrastructure, displacing fossil fuels, and reducing greenhouse gases.

Climate change concerns, coupled with high oil prices, peak oil, and increasing government support, are driving increasing renewable energy legislation, incentives and commercialization. New government spending, regulation and policies helped the industry weather the global financial crisis better than many other sectors. According to a 2011 projection by the International Energy Agency, solar power generators may produce most of the world's electricity within 50 years, dramatically reducing the emissions of greenhouse gases that harm the environment.

Growth of Renewables

From the end of 2004, worldwide renewable energy capacity grew at rates of 10–60% annually for many technologies. For wind power and many other renewable technologies, growth accelerated in 2009 relative to the previous four years. More wind power capacity was added during 2009 than any other renewable technology. However, grid-connected PV increased the fastest of all renewables technologies, with a 60% annual average growth rate. In 2010, renewable power constituted about a third of the newly built power generation capacities. By 2014 the installed capacity of photovoltaics will likely exceed that of wind, but due to the lower capacity factor of solar, the energy generated from photovoltaics is not expected to exceed that of wind until 2015.

Projections vary, but scientists have advanced a plan to power 100% of the world's energy with wind, hydroelectric, and solar power by the year 2030.

According to a 2011 projection by the International Energy Agency, solar power generators may produce most of the world's electricity within 50 years, dramatically reducing the emissions of greenhouse gases that harm the environment. Cedric Philibert, senior analyst in the renewable energy division at the IEA said: "Photovoltaic and solar-thermal plants may meet most of the world's demand for electricity by 2060 — and half of all energy needs — with wind, hydropower and biomass plants supplying much of the remaining generation". "Photovoltaic and concentrated solar power together can become the major source of electricity," Philibert said.

All forms of energy are expensive, but as time progresses, renewable energy generally gets cheaper, while fossil fuels generally get more expensive. A 2011 IEA report said: "A portfolio of renewable energy technologies is becoming cost-competitive in an increasingly broad range of circumstances, in some cases providing investment opportunities without the need for specific economic support," and added that "cost reductions in critical technologies, such as wind and solar, are set to continue."

The International Solar Energy Society argues that renewable energy technologies and economics will continue to improve with time, and that they are "sufficiently advanced at present to allow for major penetrations of renewable energy into the mainstream energy and societal infrastructures".

Barriers to Renewable Energy

The need for enacting policies to support renewable energy is often attributed to a variety of "barriers" or conditions that prevent investments from occurring. Often the result of barriers is to put renewable energy at an economic, regulatory, or institutional disadvantage relative to other forms of energy supply. Barriers include subsidies for conventional forms of energy, high initial capital costs coupled with lack of fuel-price risk assessment, imperfect capital markets, lack of skills or information, poor market acceptance, technology prejudice, financing risks and uncertainties, high transactions costs, and a variety of regulatory and institutional factors. Many of these barriers could be considered "market distortions" that unfairly discriminate against renewable energy, while others have the effect of increasing the costs of renewable energy relative to the alternatives.

Costs and Pricing

Many argue that renewable energy "costs more" than other energy sources, resulting in cost-driven decisions and policies that avoid renewable energy. In practice, a variety of factors can distort the comparison. For example, public subsidies may lower the costs of competing fuels. Although it is true that initial capital costs for renewable energy technologies are often higher on a cost-per-unit basis, it is widely accepted that a true comparison must be made on the basis of total "lifecycle" costs. Lifecycle costs account for initial capital costs, future fuel costs, future operation and maintenance costs, decommissioning costs, and equipment lifetime. Here lies part of the problem in making comparisons: What are fuel costs going to be in the future? How should future costs be discounted to allow comparison with present costs based on expected interest rates? The uncertainties inherent in these questions affect cost comparisons. Existing analytical tools for calculating and comparing costs can discriminate against renewable energy if they do not account for future uncertainties or make unrealistic assumptions.

Many policies attempt to compensate for cost-related barriers by providing additional subsidies for renewable energy in the form of tax credits or incentives, by establishing special pricing and power-purchasing rules, and by lowering transaction costs. Despite many calls for reducing subsidies for fossil fuels and nuclear power, in practice this proves politically difficult. Thus practical policies have tended to focus on increasing subsidies for renewable energy rather than reducing subsidies for fossil fuels and nuclear power.

1. *Subsidies for competing fuels.* Large public subsidies, both implicit and explicit, are channeled in varying amounts to all forms of energy, which can distort investment cost decisions. Organizations such as the World Bank and International Energy Agency put global annual subsidies for fossil fuels in the range of $100-200 billion, although such figures are very difficult to estimate (for comparison, the world spends some $1 trillion annually on purchases of fossil fuels). Public subsidies can take many forms: direct budgetary transfers, tax incentives, R&D spending, liability insurance, leases, land rights-of-way, waste disposal, and guarantees to mitigate project financing or fuel price risks. Large subsidies for fossil fuels can significantly lower final energy prices, putting renewable energy at a competitive disadvantage if it does not enjoy equally large subsidies.

2. *High initial capital costs.* Even though lower fuel and operating costs may make renewable energy cost-competitive on a life-cycle basis, higher initial capital costs can mean that renewable energy provides less installed capacity per initial dollar invested than conventional energy sources. Thus, renewable energy investments generally require higher amounts of financing for the same capacity. Depending on the circumstances, capital markets may demand a premium in lending rates for financing renewable energy projects because more capital is being risked up front than in conventional energy projects. Renewable energy technologies may also face high taxes and import duties. These duties may exacerbate the high first-cost considerations relative to other technologies and fuels.
3. *Difficulty of fuel price risk assessment.* Risks associated with fluctuations in future fossil-fuel prices may not be quantitatively considered in decisions about new power generation capacity because these risks are inherently difficult to assess. Historically, future fuel price risk has not been considered an important factor because future fossil fuel prices have been assumed to be relatively stable or moderately increasing. Thus, risks of severe fluctuations are often ignored. However, with greater geopolitical uncertainties and energy market deregulation has come new awareness about future fuel price risks. Renewable energy technologies avoid fuel costs (with the exception of biomass) and so avoid fuel price risk. However, this benefit, or "risk-reduction premium," is often missing from economic comparisons and analytical tools because it is difficult to quantify. Further, for some regulated utilities, fuel costs are factored into regulated power rates, so that consumers rather than utilities bear the burden of fuel price risks, and utility investment decisions are made without considering fuel price risk.
4. *Unfavorable power pricing rules.* Renewable energy sources feeding into an electric power grid may not receive full credit for the value of their power. Two factors are at work. First, renewable energy generated on distribution networks near final consumers rather than at centralized generation facilities may not require transmission and distribution (i.e., would displace power coming from a transmission line into a node of a distribution network). But utilities may only pay wholesale rates for the power, as if the generation was located far from final consumers

and required transmission and distribution. Thus, the "locational" value of the power is not captured by the producer. Second, renewable energy is often an "intermittent" source whose output level depends on the resource (i.e., wind and sun) and cannot be entirely controlled. Utilities cannot count on the power at any given time and may lower prices for it. Lower prices take two common forms: (i) a zero price for the "capacity value" of the generation (utility only pays for the "energy value"); (ii) an average price paid at peak times (when power is more valuable) which is lower than the value of the power to the utility—even though the renewable energy output may directly correspond with peak demand times and thus should be valued at peak prices.

5. *Transaction costs.* Renewable energy projects are typically smaller than conventional energy projects. Projects may require additional information not readily available, or may require additional time or attention to financing or permitting because of unfamiliarity with the technologies or uncertainties over performance. For these reasons, the transaction costs of renewable energy projects—including resource assessment, siting, permitting, planning, developing project proposals, assembling financing packages, and negotiating power-purchase contracts with utilities—may be much larger on a per-kilowatt (kW) capacity basis than for conventional power plants. Higher transaction costs are not necessarily an economic distortion in the same way as some other barriers, but simply make renewables more expensive. However, in practice some transaction costs may be unnecessarily high, for example, overly burdensome utility interconnection requirements and high utility fees for engineering reviews and inspection.

6. *Environmental externalities.* The environmental impacts of fossil fuels often result in real costs to society, in terms of human health (i.e., loss of work days, health care costs), infrastructure decay (i.e., from acid rain), declines in forests and fisheries, and perhaps ultimately, the costs associated with climate change. Dollar costs of environmental externalities are difficult to evaluate and depend on assumptions that can be subject to wide interpretation and discretion. Although environmental impacts and associated dollar costs are often included in economic comparisons between renewable and conventional energy, investors rarely include such environmental costs in the bottom line used to make decisions.

Legal and Regulatory

1. *Lack of legal framework for independent power producers.* In many countries, power utilities still control a monopoly on electricity production and distribution. In these circumstances, in the absence of a legal framework, independent power producers may not be able to invest in renewable energy facilities and sell power to the utility or to third parties under so-called "power purchase agreements." Or utilities may negotiate power purchase agreements on an individual ad-hoc basis, making it difficult for project developers to plan and finance projects on the basis of known and consistent rules.
2. *Restrictions on siting and construction.* Wind turbines, rooftop solar hot-water heaters, photovoltaic installations, and biomass combustion facilities may all face building restrictions based upon height, aesthetics, noise, or safety, particularly in urban areas. Wind turbines have faced specific environmental concerns related to siting along migratory bird paths and coastal areas. Urban planning departments or building inspectors may be unfamiliar with renewable energy technologies and may not have established procedures for dealing with siting and permitting. Competition for land use with agricultural, recreational, scenic, or development interests can also occur.
3. *Transmission access.* Utilities may not allow favorable transmission access to renewable energy producers, or may charge high prices for transmission access. Transmission access is necessary because some renewable energy resources like windy sites and biomass fuels may be located far from population centers. Transmission or distribution access is also necessary for direct third-party sales between the renewable energy producer and a final consumer. New transmission access to remote renewable energy sites may be blocked by transmission-access rulings or right-of-way disputes.
4. *Utility interconnection requirements.* Individual home or commercial systems connected to utility grids can face burdensome, inconsistent, or unclear utility interconnection requirements. Lack of uniform requirements can add to transaction costs. Safety and power-quality risk from non-utility generation is a legitimate concern of utilities, but a utility may tend to set interconnection requirements that go beyond what is necessary or practical for small producers, in the absence of any incentive to set more reasonable but still technically sound

requirements. In turn, the transaction costs of hiring legal and technical experts to understand and comply with interconnection requirements may be significant. Policies that create sound and uniform interconnection standards can reduce interconnection hurdles and costs.

5. *Liability insurance requirements.* Small power generators (particularly home PV systems feeding into the utility grid under "net metering" provisions) may face excessive requirements for liability insurance. The phenomenon of "islanding," which occurs when a self-generator continues to feed power into the grid when power flow from the central utility source has been interrupted, can result in serious injury or death to utility repair crews. Although proper equipment standards can prevent islanding, liability is still an issue. Several U.S. states have prohibited utilities from requiring additional insurance beyond normal homeowner liability coverage as part of net metering statutes.

Market Performance

1. *Lack of access to credit.* Consumers or project developers may lack access to credit to purchase or invest in renewable energy because of lack of collateral, poor creditworthiness, or distorted capital markets. In rural areas, "microcredit" lending for household-scale renewable energy systems may not exist. Available loan terms may be too short relative to the equipment or investment lifetime. In some countries, power project developers have difficulty obtaining bank financing because of uncertainty as to whether utilities will continue to honor long-term power purchase agreements to buy the power.
2. *Perceived technology performance uncertainty and risk.* Proven, cost-effective technologies may still be perceived as risky if there is little experience with them in a new application or region. The lack of visible installations and familiarity with renewable energy technologies can lead to perceptions of greater technical risk than for conventional energy sources. These perceptions may increase required rates of return, result in less capital availability, or place more stringent requirements on technology selection and resource assessment. "Lack of utility acceptance" is a phrase used to describe the historical biases and prejudices on the part of traditional electric power utilities. Utilities may be hesitant to develop, acquire, and maintain unfamiliar technologies, or give them proper attention in planning frameworks.

Finally, prejudice may exist because of poor past performance that is out of step with current performance norms.

3. *Lack of technical or commercial skills and information.* Markets function best when everyone has low-cost access to good information and the requisite skills. But in specific markets, skilled personnel who can install, operate, and maintain renewable energy technologies may not exist in large numbers. Project developers may lack sufficient technical, financial, and business development skills. Consumers, managers, engineers, architects, lenders, or planners may lack information about renewable energy technology characteristics, economic and financial costs and benefits, geographical resources, operating experience, maintenance requirements, sources of finance, and installation services. The lack of skills and information may increase perceived uncertainties and block decisions.

Renewable Energy Promotion Policies

Policies whose specific goal is to promote renewable energy fall into three main categories: (a) price-setting and quantity-forcing policies, which mandate prices or quantities; (b) investment cost reduction policies, which provide incentives in the form of lower investment costs; and (c) public investments and market facilitation activities, which offer a wide range of public policies that reduce market barriers and facilitate or accelerate renewable energy markets. Historically, governments have enacted these policies in a rather ad-hoc manner. More recently, national renewable energy targets (also referred to as goals) have emerged as a political context for promoting specific combinations of policies from all three categories. Such targets focus on the aggregate energy production of an entire country or group of countries. Targets may specify total primary energy from renewables and/or minimum renewable energy shares of electricity generation.

Several countries have adopted or are proposing national renewable energy targets. The European Union collectively has adopted a target of 22% of total electricity generation from renewables by 2010, with individual member states having individual targets above or below that amount. Japan has adopted a target of 3% of total primary energy by 2010. Recent legislative proposals in the United States would require 10% of electricity generation from renewables by 2020. China and India are the first developing countries to propose renewable energy targets. India has proposed that by 2012, 10% of annual additions to power generation would be from renewable energy; China has a similar goal of 5% by 2010. Other countries with existing or proposed targets are Australia, Brazil, Malaysia, and Thailand. In addition, countries from around the world placed increased attention on renewable energy targets at the U.N. World Summit for Sustainable Development in 2002.

Price-Setting and Quantity-Forcing Policies

Price-setting policies reduce cost- and pricing-related barriers by establishing favorable pricing regimes for renewable energy relative to other sources of power generation. The quantity of investment obtained under such regimes is unspecified, but prices are known in advance. Quantity-forcing policies do the opposite; they mandate a certain percentage or absolute quantity of generation to be supplied from renewable energy, at unspecified prices. Often price-setting or quantity-forcing policies occur in parallel with other policies, such as investment cost reduction policies.

The two main price-setting policies seen to-date are the PURPA legislation in the United States and "electricity feed-in laws" in Europe. The two main quantity-forcing policies seen to-date are competitively-bid renewable-resource obligations and renewable portfolio standards.

U.S. Public Utility Regulatory Policies Act (PURPA). PURPA was enacted in 1978 in part to encourage electric power production by small power producers using renewable resources to reduce U.S. dependence on foreign oil. The policy required utilities to purchase power from small renewable generators and cogenerators, known as "qualifying facilities," through long-term (10-year) contracts at prices approximating the "avoided costs" of the utilities. These avoided costs represented the marginal costs to the utilities of building new generation facilities, which could be avoided by purchasing power from the qualifying facilities instead. Avoided cost calculations typically assumed an aggressive schedule of escalating future energy prices, making contract prices to qualifying facilities quite attractive.

For example, "standard offer" contracts under PURPA in California in the 1980's set the avoided cost of generation in the range 6-10 cents/kWh—very favorable rates which initially spurred many investments by renewable energy producers. However, by the 1990s energy prices had not risen as originally expected and a large number of natural-gas fired independent generation came on-line in California.Power surpluses emerged, wholesale power prices declined, and declining standard offer rates led to reduced competitiveness of renewable energy and a significant slowdown in construction of new capacity.

Electricity Feed-in Laws. The electricity feed-in laws in Germany, and similar policies in other European countries in the 1990s, set a fixed price for utility purchases of renewable energy. For example, in Germany starting in 1991, renewable energy producers could sell their power to utilities at 90% of the retail market price. The utilities were obligated to purchase the power. The German feed-in law led to a rapid increase in installed capacity and development of commercial renewable energy markets. Wind power purchase prices were highly favorable, amounting to about DM 0.17/kWh (US 10 cents/kWh), and applied over the entire life of the plant. Total wind power installed went from near zero in the early 1990s to over 8500 MW by 2001, making Germany the global leader in renewable energy investment.

Partly because retail electricity prices declined with increasing competition due to electricity deregulation, which made producers and financiers wary of new investments, a new German Renewable Energy Law of 2000 changed electricity feed-in pricing. Pricing became based on fixed norms unique to each technology, which in turn were based upon estimates of power production costs and expectations of declines in those costs over time. For example, wind power prices remained at the previous level of DM 0.17/kWh for plants commissioned in 2001, but only for the first five years of operation, after which prices paid declined. Solar PV prices were set initially at DM 0.99/kWh. All prices had build-in declines over time (i.e., 1.5% annual decreases in starting tariffs paid for wind power plants commissioned in subsequent years). This provision addressed one of the historical criticisms of feed-in approaches, which was that they did not encourage technology cost reductions or innovation. The new law's provisions for regular adjustments to prices addressed technological and market developments. The law also distributed the costs of the policy (i.e., the additional costs of wind power over conventional power) among all utility customers in the country. This issue of burden sharing had become a significant political issue in Germany by 2000 because the old law placed a disproportionate burden on utility customers in specific regions where wind power development was heaviest.

Other countries in Europe with renewable electricity feed-in laws include Denmark, France, Greece, Italy, Portugal, Spain, and Sweden. A combination of feed-in tariffs, production subsidies of DK 0.10/kWh, and a strong domestic market helped the Danish wind industry maintain a 50% market share of global wind turbine production for a number of years.

Competitively-bid renewable-resource obligations. The United Kingdom tried competitive bidding for renewable-energy-resource obligations during the 1990s under its "Non-Fossil-Fuel Obligation" (NFFO) policy. Under the NFFO, power producers bid on providing a fixed quantity of renewable power, with the lowest-price bidder winning the contact. With each successive bidding round (there were four total), bidders reduced prices relative to the last round. For example, wind power contract prices declined from 10 p/kWh in 1990 under NFFO-1, to 4.5 p/kWh in 1997 under NFFO-4. One of the lessons some have drawn from the UK experience is that competitively determined subsidies can lead to rapidly declining prices for renewable energy. However, there has also been criticism that the NFFO process encouraged competing projects to bid below cost in order to capture contracts, with the result that successful bidders were unable to meet the terms of the bid or ended up insolvent. This criticism proved valid in practice; contracts awarded to low-bidders did not always translate into projects on the ground. The UK abandoned the NFFO approach after the fourth round of bidding in 1997. Other countries with similar competitively-bid renewable resource mechanisms have included Ireland (under the "AER" program), France (under the "EOLE" program), and Australia (under the "RECP" program).

Renewable energy portfolio standards (RPS). An RPS requires that a minimum percentage of generation sold or capacity installed be provided by renewable energy. Obligated utilities are required to ensure that the target is met, either through their own generation, power purchases from other producers, or direct sales from third-parties to the utility's customers. Typically, RPS obligations are placed on the final retailers of power, who must purchase either a portion of renewable power or the equivalent amount of green certificates. Two types of standards have emerged: *capacity-based standards* set a fixed amount of capacity by a given date, while *generation-based standards* mandate a given percentage of electricity generation that must come from renewable energy.

In the United States, many RPS policies have occurred as part of utility restructuring legislation. These are typically generation-based standards with phased implementation to allow utilities to reach incrementally increasing targets over a number of years. At least twelve U.S. states have enacted an RPS, ranging from 1% to 30% of electricity generation. However, the amount of *new and additional* generation expected from these standards varies widely depending on existing renewable energy capacity. For example, Maine historically generates over 40% of its power from renewable resources, so its 30% standard is unlikely to result in any new renewable generation. In contrast, California's requirement to increase renewable sales from 10.5% in 2001 to 20% by 2017 will likely result in a significant amount of new in-state renewable energy generation. Texas implemented an RPS in 2000 requiring 2000 MW of new renewable capacity by 2008. Partly due to federal production tax credits, Texas has been substantially ahead of schedule, with half of the targeted capacity in place by 2002.

In Europe, the Netherlands has been a leader among RPS initiatives. Dutch utilities have adopted an RPS voluntarily, based on targets of 5% of electricity generation by 2010, increasing to 17% by 2020. Other countries with regulatory requirements for utilities or electricity retailers to purchase a percentage of renewable power include Australia, Brazil, Belgium, Denmark, France, Japan, Spain, Sweden, and the United Kingdom. The UK's 'Renewables Obligation' on suppliers will rise in annual steps from 3% in 2003 to 10% in 2010. In Denmark, legislation obliges end users to purchase 20% of their electricity from renewable sources by 2003.

In Brazil, a policy enacted during the electricity crisis of 2001 requires national utilities to purchase over 3000 MW of renewable energy capacity by 2016. Purchase prices are set by the government at 80 percent of the national average electricity retail price. Thus, in contrast to most national policies elsewhere, Brazil's policy is effectively both "price-setting" and "quantity-forcing."

Renewable Energy (Green) Certificates. Renewable energy (green) certificates are emerging as a way for utilities and customers to trade renewable energy production and/or consumption credits in order to meet obligations under RPS and similar policies. Standardized certificates provide evidence of renewable energy production, and are coupled with institutions and rules for trading that separate renewable attributes from the associated physical energy. This enables a "paper" market for renewable energy to be created independent of actual electricity sales and flows. Green certificate markets are emerging in several countries, allowing producers or purchasers of renewable energy who earn green certificates to sell those certificates to those who need to meet obligations but haven't generated or purchased the renewable power themselves. Those without obligations, but wishing to voluntarily support green power for philosophical or public-relations reasons may also purchase certificates.

Public and private institutions are emerging that keep track of renewable energy generation, assign certificates to generators, and register trades and sales of certificates. Green certificate trading is gaining ground in the UK, Belgium, Denmark, Australia, and the United States. Europe embarked upon a "test phase" of an EU-wide renewable energy certificate trading system during 2001 and 2002; more than 40 companies in 7 countries had opened trading accounts, with a further 100 companies intending to join the test phase in late 2002. Certificates for over 1,000 GWh were issued through 2002.

Cost Reduction Policies

A number of policies are designed to provide incentives for voluntary investments in renewable energy by reducing the costs of such investments. These policies can be characterized as falling in five broad categories: policies that (1) reduce capital costs up front (via subsidies and rebates); (2) reduce capital costs after purchase (via tax relief); (3) offset costs through a stream of payments based on power production (via production tax credits); (4) provide concessionary loans and other financial assistance, and (5) reduce capital and installation costs through economies of bulk procurement.

Subsidies and Rebates. Reduction in the initial capital outlay by consumers for renewable energy systems is accomplished through direct subsidies, or rebates. These subsidies are used to "buy down" the initial capital cost of the system, so that the consumer sees a lower price. In the United States, at least nineteen states offer rebate programs at the state, local, and/or utility level to promote the installation of renewable energy equipment. The majority of the programs are available from state agencies and municipally-owned utilities and support solar water heating and/or photovoltaic systems, though some include geothermal heat pumps, small wind generators, passive solar, biomass, and fuel cells. Homes and businesses are usually eligible, although some programs target industry and public institutions as well. In some cases, rebate programs are combined with low or no-interest loans.

Sustained efforts to increase the use of renewables have been made via coordinated, multi-year, multi-policy initiatives. For example, Japan, Germany, and the United States subsidize capital costs of solar PV as part of their "market transformation" programs.

- Japan's Sunshine Program provides capital subsidies and net metering for rooftop PV systems. From 1994 to 2000, the government invested 86 billion yen ($725 million USD), resulting in 58,000 system installations and over 220 MW of PV capacity. Subsidies began at 900,000 yen/peak-kW (US$5/peak-watt) in 1994, and were gradually reduced to 120,000 yen/peak-kW (US$1/peak-watt) in 2001 as PV prices fell.
- Germany began a "1000 solar roofs" program in 1991 that offered subsidies for individual household purchases of solar PV of up to 60% of capital system costs. The program was expanded in 1999 to 100,000 roofs over five years, providing 10-year low-interest loans to households and businesses. As a result of favorable feed-in tariffs and low-interest loans, the program was expected to provide 300 peak-MW of PV capacity.
- The United States launched a "million solar roofs" initiative in 1997 to install solar PV systems and solar thermal systems on one million buildings by 2010. The program includes long-term low-interest customer financing, government procurement for federal buildings, commercialization programs, and production incentives. Individual states also have capital subsidy programs for PV, with the California Energy Commission offering rebates of up to $4.50/peak-watt or 50 percent off the system purchase price, New York and New Jersey offering up to $5/peak-watt subsidies, and New York rebating as much as 70% of the cost of eligible equipment as of 2002.

Subsidy programs also exist for wind power, such as Denmark's DK 0.10/kWh (US 1.5 cents/kWh) production subsidy paid to utilities. Among developing countries, Thailand provided subsidies for small renewable energy power producers starting in 2000, soliciting bids for 300 MW of small renewable power, and providing production subsidies above standard power purchase rates for at least the first five years of operation of each facility.

Tax Relief. Tax relief policies to promote renewable energy have been employed in the United States, Europe, Japan, and India. Tax relief has been especially popular in the United States, where a host of federal and state tax policies address energy production, property investments, accelerated depreciation, and renewable fuels. State policies vary widely in scope and implementation. At least 17 states have personal tax incentives, 21 states have corporate tax incentives, 16 states have sales tax incentives, and 24 states have property tax incentives.

a. *Investment Tax Credits.* Investment tax credits for renewable energy have been offered for businesses and residences. In the United States, businesses receive a 10 percent tax credit for purchases of solar and geothermal renewable energy property, subject to certain limitations. Some U.S. states have investment tax credits of up to 35%.

b. *Accelerated Depreciation.* Accelerated depreciation allows renewable energy investors to receive the tax benefits sooner than under standard depreciation rules. The effect of accelerated depreciation is similar to that of investment tax credits. In the United States, businesses can recover investments in solar, wind, and geothermal property by depreciating them over a period of five years, rather than the 15- to 20-year depreciation lives of conventional power investments.

India's accelerated depreciation policy allowed 100% depreciation in the first year of operation, helping spur the largest wind power industry among developing countries. However, this policy led to large investments without sufficient regard to long-term operating performance and maintenance, resulting in capacity factors lower than for wind power installations elsewhere. This led many to conclude that production-based incentives are preferable to investment tax credits and accelerated depreciation, although Germany's investment tax credits accompanied by wind turbine technical standards and certification requirements avoided the problems found in India.

c. *Production Tax Credits.* A production tax credit provides the investor or owner of qualifying property with an annual tax credit based on the amount of electricity generated by that facility. By rewarding production, these tax credits encourage improved operating performance. A production tax credit in Denmark provides DK 0.10/kWh (US 1.5 cents/kWh) for wind power, but few other countries have adopted similar credits. In the United States the Renewable Electricity

Production Credit (PTC) provides a per-kWh tax credit for electricity generated by qualified wind, closed-loop biomass, or poultry waste resources. Federal tax credits of 1.5 cents/kWh (adjusted annually for inflation) are provided for the first ten years of operation for all qualifying plants that entered service from 1992 through mid-1999, later extended to 2001 and then to 2003.

At least five U.S. states have state or local production incentives for distributed electrical generation, renewable fuels, or both. These policies are similar to the federal PTC, with specific limits on technologies, dates-in-service, and maximum payout per provider and per year. Funds to support the incentives are obtained from a mixture of sources, including general funds, public benefit or environmental funds, and green electricity sales (so-called "green tags").

d. *Property Tax Incentives.* At least 24 U.S. states have property tax incentives for renewable energy. These incentives are implemented on many scales—state, county, city, town, and municipality. These are generally implemented in one of three ways: (1) renewable energy property is partially or fully excluded from property tax assessment, (2) renewable energy property value is capped at the value of an equivalent conventional energy system providing the same service, and (3) tax credits are awarded to offset property taxes.

e. *Personal Income Tax Incentives.* Credits against personal state income taxes are available in many U.S. states for purchase of and/or conversion to eligible renewable energy systems and renewable fuels. In some cases, taxpayers can deduct the interest paid on loans for renewable energy equipment.

f. *Sales Tax Incentives.* At least sixteen U.S. states have policies that provide retail sales tax exemptions for eligible renewable energy systems and renewable fuels. Most exempt 100% of the sales tax for capital expenses, and provide specific cents-per-gallon exemptions for renewable fuels. Some policies specify maximum or minimum sizes for eligible systems.

g. *Pollution Tax Exemptions.* The Netherlands is an example where "green" power is exempt from a new and rising fossil-fuel tax on electricity generation that is paid by end-users. Starting in 2001, that fossil-fuel tax rose to the equivalent of US 5 cents/kWh, providing a large tax incentive for Dutch consumers.

h. *Other Tax Policies.* A variety of other tax policies exist, such as income tax exemptions on income from renewable power production, excise duty and sales tax exemptions on equipment purchased, and reduced or zero import tax duties on assembled renewable energy equipment or on components. India, for example, has allowed five-year tax exemptions on income from sales of wind power.

Grants. Many countries have offered grants for renewable energy purchases. For example, beginning in 1979, Denmark provided rebates of up to 30% of capital costs for wind and other renewable energy technologies. These rebates declined over time. In the United States, county and state governments and utilities provide grants for renewable energy ranging in size from hundreds to millions of dollars.

Loans. Loan programs offer financing for the purchase of renewable energy equipment. Loans can be market-rate, low-interest (below market rate), or forgivable. In many U.S. states, loans are available to virtually all sectors—residential, commercial, industrial, transportation, public, and nonprofit. Repayment schedules vary, with terms of up to 10 years common. Interest rates for renewable energy investments can often be 1% or more higher than those for conventional power projects because of the higher perceived risks involved, so government-subsidized loans that offer below-market interest rates are also common.

Renewable energy loans can take many forms. Residential loans may range from $500 to $10,000 or more, while commercial and industrial loans may extend to the millions. Funding comes from a variety of sources, including municipal bonds, system benefit funds, revolving funds, and utility penalty or overcharge funds. Financing may be for a fraction to 100% of a project. Some loan programs have minimum or maximum limits, while others are open-ended. Loan terms range from 3 years to the life of a project. Some loans are contractor-driven, and may include service contracts in the loan amount. Sometimes grants and loans are combined; for example, Iowa provides a 20% forgivable loan combined with an 80% loan at prime rate for renewable fuels projects.

In some developing countries, notably India, China, and Sri Lanka, multilateral loans by lenders such as the World Bank have provided financing for renewable energy, usually in conjunction with commercial lending. One of the most prominent examples is the India Renewable Energy Development

Agency (IREDA), which was formed in 1987 to provide assistance in obtaining international multilateral agency loans and in helping private power investors obtain commercial loans. By 2001, IREDA had disbursed the equivalent of over US$400 million in loans for renewable energy projects in India, resulting in over 1600 MW of renewable power generation.

Public Investments and Market Facilitation Activities

1. *Public Benefit Funds.* In the United States, public funds for renewable energy development are raised through a System Benefits Charge (SBC), which is a per-kWh levy on electric power consumption. Some analysts suggest that state clean energy funds seem to be one of the more effective policies in promoting renewable energy development to result from electricity restructuring. It is estimated that fourteen U.S. states will collect $3.5 billion through 2011 in system benefits charges. Similar levies exist in some European countries for fossil-fuel-based generation. In general, the funds serve a variety of purposes, such as paying for the difference between the cost of renewables and traditional generating facilities, reducing the cost of loans for renewable facilities, providing energy efficiency services, funding public education on energy-related issues, providing low-income energy assistance, and supporting research and development.
2. *Infrastructure Policies.* Market facilitation supports market institutions, participants, and rules to encourage renewable energy technology deployment. A variety of policies are used to build and maintain this "market infrastructure," including policies for design standards, accelerated siting and permitting, equipment standards, and contractor education and licensing. Additionally, policies to induce renewable technology manufactures to site locally, and direct sales of renewable systems to customers at concessionary rates facilitate market development.
    a. *Construction and design policies.* Construction and design standards include building-code standards for PV installations, design standards evaluated on life-cycle cost basis, and performance requirements. Policy examples include Tucson, Arizona, which requires that commercial facilities achieve a 50% reduction in energy usage over 1995 Model Energy Code, and Florida, which requires that all new educational facilities include passive solar design.

b. *Site prospecting, review and permitting.* Federal and state programs reduce barriers to renewable energy development through resource, transmission, zoning, and permitting assessments. This particularly helped early promotion of wind energy projects in California. On a national scale the Utility Wind Resource Assessment Program funds a number of supporting activities, including up to 50% of the cost of wind resource assessments. India also has a large wind assessment program, with over 600 stations in 25 states providing information to project developers on the best sites for development.

c *Equipment standards and contractor certification.* A variety of equipment-related standards and certification measures have been applied to ensure uniform quality of equipment and installation, increasing the likelihood of positive returns from renewable energy installations. Contractor licensing requirements ensure that contractors have the necessary experience and knowledge to properly install systems. Equipment certifications, ensure that equipment meets certain minimum standards of performance or safety.

d. *Industrial recruitment.* Industrial recruitment policies use financial incentives such as tax credits, grants, and government procurement commitments to attract renewable energy equipment manufacturers to locate in a particular area. These incentives are designed to create local jobs, strengthening the local economy and tax base, and improving the economics of local renewable development initiatives.

e. *Direct equipment sales.* These programs allow the consumer to buy or lease renewable energy systems directly from electric provider at below-retail rates. Some programs provide a capital buydown. Examples include Arizona, which provides a buydown of $2/peak-watt for PV, and California's SMUD, which offers a 50% buydown plus 10-year financed loans and net metering.

3. *Government Procurement.* Government procurement policies aim to promote sustained and orderly commercial development of renewable energy. Governmental purchase agreements can reduce uncertainty and spur market development through long-term contracts, pre-approved purchasing agreements, and volume purchases. Government purchases of renewable energy technologies in early market stages can help overcome institutional barriers to commercialization, encourage the development of appropriate infrastructure, and provide a "market path" for technologies that require integrated technical, infrastructure, and regulatory changes.

4. *Customer Education and Mandated Generation Disclosure Information.* U.S. restructuring and deregulation policies mandate that information be provided to customers about choice of electricity providers and "characteristics" of electricity being provided (such as emissions and fuel types). In many states, general education to raise customer awareness about renewable energy and the environmental impacts of energy generation is required, typically via websites and printed materials.
5. *Solar and Wind Access Laws.* Renewable access laws address access, easements, and covenants. Access laws provide a property owner the right to continued access to a renewable resource. Easements provide a privilege to have continued access to wind or sunlight, even though development or features of another person's property could reduce that access. Easements are often voluntary contracts, and may be transferred with the property title. Covenant laws prohibit neighborhood covenants from explicitly restricting the installation or use of renewable equipment. Policy mechanisms include access ordinances, development guidelines addressing street orientation, zoning ordinances with building height restrictions, and renewable permits.

Transport Biofuels Policies

Biofuels mandates and tax policies in Brazil, the United States, and Europe have supported accelerating development of biofuels. Biofuels mandates require that a certain percentage of all liquid transport fuels be derived from renewable resources. Tax policies may provide tax credits or exemptions for production or purchase of biofuels.

Brazil has been quite successful with biofuels mandates under its "ProAlcool" program, which has promoted the use ethanol for transportation fuel since the 1980s. In addition to a variety of economic incentives and subsidies, Brazil has mandated that ethanol be blended with all gasoline sold in the country. Brazil has also required that all gas stations sell pure ethanol. This last requirement made it commercially viable for the automotive industry to produce ethanol-only (neat ethanol) cars. However, the share of ethanol-only cars purchased annually, after rising to 95% by 1985, subsequently declined for a number of reasons. Today, most cars use the "gasohol" ethanol/gasoline blend, and more than 60 percent of Brazil's sugar cane production goes to produce 18 billion gallons of ethanol each year, representing 90% of global ethanol production. In 2000, over 40% of automobile fuel consumption in Brazil was ethanol.

The United States, the world's second largest ethanol producer after Brazil, has a number of biofuels tax policies and mandates. The Energy Security Act of 1979 created a federal ethanol tax credit of up to 60 cents per gallon for businesses that sell or use alcohol as a fuel. Gasoline refiners and distributors may also receive an excise tax exemption of up to 5 cents/gallon for blending their fuels with ethanol. State-level ethanol policies also exist, whose origins in the 1980s can be traced back to initiatives by Iowa to use its corn crop for energy. Several policies in Iowa were established to encourage ethanol consumption, including a mandate for government vehicles to use ethanol-blended fuel, and a one-cent-per-gallon fuel sales tax exemption for ethanol-blended fuels. In 1998, both the federal government and the State of Iowa extended their ethanol tax exemptions until the year 2007. In part due to ethanol incentives, over sixty ethanol production facilities have become operational in the U.S. since 1976, with a production capacity of more than 2.4 billion gallons per year. A recent U.S. legislative proposal, called a "renewable fuels standard," would triple biofuel use within 10 years.

In Europe, Germany is the largest user of biodiesel and provides tax incentives for 100% pure biodiesel. These incentives have had a large effect; German consumption of biodiesel went from 200 million gallons in 1991 to 750 million gallons in 2002. Other EU members provide tax incentives for 2-5% biodiesel blends. Germany, Austria, and Sweden use 100% pure biodiesel in specially adapted vehicles, and biodiesel is mandated in environmentally sensitive areas in Germany and Austria. Other countries producing biodiesel are Belgium, France, and Italy, the later two also providing tax incentives. The European Commission has recently proposed a biofuel directive with targets or mandates for biofuels up to 6% of all transportation fuel sold.

Among other developing countries, Thailand is considering tax policies for biofuels, including excise tax exemptions for ethanol and income tax waivers for investments in biofuels facilities. Malaysia and Indonesia also utilize biodiesel.

Emissions Reduction Policies

Policies to reduce power plant emissions, including NO_x, SO_x, and CO_2, have the potential to affect renewable energy development. Many emissions-reduction policies create "allowances" for certain emissions (representing the right to emit a certain amount of that pollutant).Credits available to renewable energy generation can "offset" these allowed emissions. Such credits have market value, and are often traded to allow electricity generators to comply with emission regulations at least cost. Three innovative examples of U.S. air quality standards, acid rain prevention programs, and state-level greenhouse-gas reduction initiatives illustrate the potential of emissions reductions policies to affect renewable energy.

Renewable Energy Set-Asides

To meet National Ambient Air Quality Standards, the U.S. Environmental Protection Agency requires twenty-two U.S. states and the District of Columbia to reduce NO_x emissions significantly by 2007. States can meet emission reduction targets through actual emission reductions and/or purchase of emission reduction credits from other states participating in a region-wide NO_x trading program. States can allocate, or "set-aside," a percentage of the total state NO_x allowances to energy efficiency and renewable energy. Eligible renewable energy producers receive these set-aside allowances and can sell them to fossil-fuel-based electricity generators to enable those generators to stay within their NO_x allocation. The additional revenue from sales of these set-aside allowances can potentially provide stimulus for renewable energy development, although to date few states have implemented renewable energy set-asides.

Emissions Cap-and-Trade Policies

Under the 1990 Clean Air Act the United States instituted a cap-and-trade mechanism to reduce SO_2 emissions and facilitate least-cost compliance. Under the Acid Rain sub-program, 300,000 SO_2 emissions allowances (rights to emit SO_2) were set aside for utilities that employed renewable energy or energy efficiency measures. Allowances were to be earned from 1992 through 1999, allocated at a rate of one allowance (one ton of SO_2 avoided) per 500 MWh of generation produced by renewable energy or avoided through increased energy efficiency. This program was not particularly effective, as only one-tenth of the 300,000 allowances were allocated to energy efficiency or renewable energy. Analysis suggests that allowance prices set by the market were not high enough to justify renewable energy investments. In addition, the program restricted participation to utilities, excluding other power generators, which also contributed to under-subscription.

Greenhouse Gas Mitigation Policies

The New Jersey Sustainability Greenhouse Gas (GHG) Action Plan was designed to promote the capture of landfill methane gas for power generation, thus avoiding methane emissions (methane has a high "global warming potential"). A loan fund provides New Jersey companies with low-interest loans to support initiatives including development of biomass/landfill methane energy resources, and emission offsets from landfill methane and other renewable power generation could potentially be included in GHG trading systems.

Power Sector Restructuring Policies

Power sector restructuring is having a profound effect on electric power technologies, costs, prices, institutions, and regulatory frameworks. Restructuring trends are changing the traditional mission and mandates of electric utilities in complex ways, and affecting environmental, social, and political conditions. There are five key trends underway that continue to influence renewable energy development, both positively and negatively, as discussed below.

Power Markets and Removal of Price Regulation on Generation

Power generation is usually one of the first aspects of utility systems to be deregulated. The trend is away from utilities monopolies towards open competition, where power contracts are signed between buyers and sellers in wholesale "power markets." Distribution utilities and industrial customers gain more choices in obtaining wholesale power. Such markets may often begin with independent-power-producer (IPP) frameworks. As wholesale electricity becomes more of a competitive market commodity, price becomes relatively more important than other factors in determining a buyer's choice of electricity supplier.

The potential effects of competitive wholesale markets and IPPs on renewable energy are significant. Wholesale power markets allow IPPs to bypass the biases against renewables that traditional utility monopolies have had. Indeed, one of the very first major markets for renewable energy in the 1980s was in California, spurred by the PURPA legislation discussed earlier. In some countries, IPP frameworks have been explicitly enacted to support renewable energy. Examples are Sri Lanka and Thailand, where utility monopolies were broken and renewable energy IPPs can sell power to the grid. However, other effects of wholesale competition may stifle renewable energy development. As low-cost combined-cycle gas turbines begin to dominate new generation, renewable energy has difficulty competing on the basis of price alone. In addition, the emergence of short-term power contracts and 'spot' markets favor generation technologies with higher variable costs and lower capital costs, like fossil fuels, rather than capital-intensive but low-operating-cost technologies like renewables.

Self-Generation by End-Users and Distributed Generation Technologies

Independent power producers may be the end-users themselves rather than just dedicated generation companies. With the advent of IPP frameworks, utility buy-back schemes (including net metering), and cogeneration technology options, more and more end-users, from large industrial customers to small residential users, are generating their own electricity. Their self-generation offsets purchased power and they may even sell surplus power back to the grid. Traditionally, regulated monopoly utilities have enjoyed economic advantages from large power plants and increasing economies of scale. These advantages are eroding due to new distributed generation technologies that are cost-competitive and even more efficient at increasingly smaller scales. In fact, newer technologies reduce investment risks and costs at smaller scales by providing modular and rapid capacity increments.

Renewable energy is well suited to self-generation, but faces competition from other distributed generation technologies, especially those based on natural gas. Gas has become the fuel of choice for small self-producers because of short construction lead times, low fuel and maintenance costs, and modular small-scale technology. However, with restructuring, a host of distributed generation policies, including net metering, become possible. These policies often spur renewable energy investments. On the other hand, self-generators may be penalized by utility-wide surcharges that accompany restructuring, such as those for stranded generation assets (called "non-bypassable competitive transition charges" in the U.S.). Self-generators who use renewable energy must still pay these charges, based on the amount of electricity they *would* have purchased from the grid, even if actual grid consumption is small.

Privatization and/or Commercialization of Utilities

In many countries, utilities, historically government-owned and operated, are becoming private for-profit entities that must act like commercial corporations. Even if utilities remain state-owned, they are becoming "commercialized"—losing state subsidies and becoming subject to the same tax laws and accounting rules as private firms. In both cases, staffing may be reduced and management must make independent decisions on the basis of profitability.

The effects of privatization and commercialization on renewable energy are difficult to judge. The environmental effects of privatization can be positive or negative, depending on such factors as the strength of the regulatory body and the political and environmental policy situation in a country. Private utilities are more likely to focus more on costs and less on public benefits, unless specific public mandates exist. On the positive side, privatization may promote capital-intensive renewable energy by providing a new source of finance—capital from private debt and equity markets. However, the transition from public to private may shorten time horizons, increase borrowing costs, and increase requirements for high rates of return. All of these factors would limit investments in capital-intensive renewable energy projects, in favor of lower-capital-cost, higher-operating-cost fossil-fuel technologies.

Unbundling of Generation, Transmission and Distribution

Utilities have traditionally been vertically integrated, including generation, transmission and distribution functions. Under some restructuring programs, each of these functions is being "unbundled" into different commercial entities, some retaining a regulated monopoly status (particularly distribution utilities) and others starting to face competition (particularly generators).

Unbundling can provide greater consumer incentives to self-generate using renewable energy. If retail tariffs are "unbundled" as well, so that generation, transmission and distribution costs are separated, customers have more incentive to self-generate, thereby avoiding transmission and distribution charges. In addition, open-access transmission policies that go along with unbundling have been explicitly targeted to promote renewable energy in some countries. In India, open-access policies helped catalyze the wind industry there, by allowed firms to produce wind power in remote regions with good wind resources and then "wheel" the power over the transmission system to their own facilities or to third parties. Brazil enacted a 50% reduction of transmission wheeling fees for renewable energy producers, which has been credited with promoting a booming small hydro industry there. However, unbundling can also penalize inherently intermittent renewable energy if producers have to pay transmission charges on a per-capacity basis. That is, even when the transmission capacity is not being used (say the wind is not blowing), transmission charges must be paid, resulting in high average transmission costs per kWh.

Competitive Retail Power Markets and "Green Power" Sales

Competition at the retail level, the newest phenomena in power sector restructuring, means that individual consumers are free to select their power supplier from among all those operating in a given market. Competitive retail power markets have allowed the emergence of "green power" suppliers who offer to sell renewable energy, usually at a premium. As green power sales grow, these suppliers are forced to investment in new renewable energy capacity to meet demand, or buy power from other renewable energy producers. Green power markets have begun to flourish where retail competition is allowed, but often only in conjunction with other renewable energy promotion policies.

The Netherlands is perhaps the best-known example. Following restructuring in 2001, one million green power customers signed up within the first year. However, incentives played a role; a large tax on fossil-fuel generated electricity, from which green power sales were exempt, made green power economically competitive with conventional power. In the U.S., green power markets are emerging in several states in response to state incentives and aggressive marketing campaigns by green power suppliers. At least 30 U.S. states have green pricing programs. Four states have mandatory green power policies that require utilities to offer customers opportunities to support renewable energy. California became one of the largest markets, with over 200,000 customers, but this was aided by a 1 cent/kWh subsidy to green power, paid for by a system benefits charge.

Distributed Generation Policies

Distributed generation avoids some of the costs of transmission and distribution infrastructure and power losses, which together can total up to half of delivered power costs. Policies to promote distributed generation—including net metering, real-time pricing, and interconnection regulations—do not apply only to renewable energy, but nevertheless can strongly influence renewable energy investments.

Net Metering

Net metering allows a two-way flow of electricity between the electricity distribution grid and customers with their own generation. When a customer consumes more power than it generates, power flows from the grid and the meter runs forward. When a customer installation generates more power than it consumes, power flows into the grid and the meter runs backward. The customer pays only for the net amount of electricity used in each billing period, and is sometimes allowed to carryover net electricity generated from month to month. Net metering allows customers to receive retail prices for the excess electricity they generate at any given time. This encourages customers to invest in renewable energy because the retail price received for power is usually much greater than it would be if net metering were not allowed and customers had to sell excess power to the utility at wholesale rates or avoided costs. Electricity providers may also benefit from net metering programs, particularly with customer-sited PV which produces electricity during peak periods. Such peak power can offset the need for new central generation and improve system load factors. At least 38 U.S. states now have net metering laws. Size limits on net metered systems typically range from 10kW to 100kW, with the exception of a few states that do not limit system size or overall enrollment. Net metering is common in parts of Germany, Switzerland and the Netherlands, and allowed by at least one utility in the UK. Thailand is one of the few developing countries to have enacted net metering laws, following a pilot project for net-metered rooftop PV in the 1990's.

Real-Time Pricing

Real-time pricing, also known as dynamic pricing, is a utility rate structure in which the per-kWh charge varies each hour based on the utility's real-time production costs. Because peaking plants are more expensive to run than baseload plants, retail electricity rates are higher during peak times than during shoulder and off-peak times under real-time pricing. When used in conjunction with net metering, customers receive higher peak rates when selling power into the grid at peak times. At off-peak times the customer is likely purchasing power from the grid, but at the lower off-peak rate. Photovoltaic power is often a good candidate for real-time pricing, especially if maximum solar radiation occurs at peak-demand times of day when power purchase prices are higher. Real-time metering equipment is necessary, which adds complexity and expense to metering hardware and administration.

Real-time pricing has been used with some large power consumers for decades. For example, power companies in Nova Scotia and New York state offer real time pricing rates for large commercial and industrial customers that vary hourly according to the varying cost of generation. In a recent pilot project, California installed 23,000 real-time meters for large customers at a cost of $35 million. In response, summer peak demand by those customers dropped by 500 MW under time-of-use pricing, which would allow the utility to avoid $250-300 million in capacity additions. Although real-time pricing has not become widespread, with favorable rate structures it has the potential to provide significant incentives for grid-connected renewable development.

Interconnection Regulations

Non-discriminatory interconnection laws and regulations are needed to address a number of crucial barriers to interconnection of renewable energy with the grid. Interconnection regulations often apply to both distributed generation and "remote" generation with renewable energy that requires transmission access, such as wind power.

1. *Legal Access.* The ability to legally connect a renewable energy system to a grid depends on federal, state, and local government rules and regulations. These policies both allow connection and determine how physical connection is achieved. In the U.S., the legal right to connect to the grid is provided for in federal laws such as the Public Utilities Regulatory Policies Act (PURPA) of 1978, and by state net metering statutes.

2. *Dynamic Generation and Transmission Scheduling.* Historically, transmission policies have often imposed severe penalties on unscheduled deviations from projected (advance-scheduled) power generation. These penalty structures render intermittent generation, such as wind or PV, uneconomic. Real-time accounting of power transfer deviations that provides charges or credits to producers based upon the value of energy at the time of the deviation, as well as elimination of discriminatory deviation penalties, allows intermittent renewable energy to compete more equitably with traditional generation.

   Policies that allow near-time or real-time scheduling of the output levels of intermittent resources can further reduce deviation costs. For example, wind farms are able to predict their output much more accurately up to an hour in advance of generation, and thus can be better scheduled hour-by-hour rather than day-ahead.

3. *Elimination of Rate "Pancaking."* Because distributed renewables, such as wind, are often remotely located, they can incur high transmission fees as power crosses multiple jurisdictions to get to the customer. Such cumulative addition of transmission fees is known as "rate pancaking." Elimination of access rate pancaking, either by consolidation of long-distance tariffs under a regional transmission organization, and/or by creating access waiver agreements between multiple owner/operators, can reduce discrimination against wind and other remote distributed renewables.

4. *Capacity Allocation.* When demand for a transmission path exceeds its reliable capacity, transmission congestion occurs. In such circumstances, system operators must allocate available capacity among competing users. Traditional utility policies often favor early market entrants, "grandfathering" them into capacity allocation rules. Wind power is particularly susceptible to transmission constraints, as it is generally located far from load. Elimination of grandfathering would allow transmission users to bid for congested capacity on an equal-footing. Allowing wind to bid for congested capacity closer to the operating hour, and reducing congestion through transmission line upgrades would also reduce barriers to wind energy development.

5. *Standard Interconnection Agreements.* Utilities may require the same interconnection procedures for small systems as are required for large independent power production facilities. The process of negotiating a

power purchase/sale contract with the utility can be very expensive, and utilities can charge miscellaneous fees that greatly reduce the financial feasibility of small grid-connected renewable installations. Standardized interconnection agreements can expedite this process. Some believe that Texas provides a good model for renewable interconnection. Under a standard agreement, renewable developers pay only for the direct costs of connecting the plant to the local system, but not for upgrades to the grid necessary to carry additional capacity. This allows generators to compete more equally.

Rural Electrification Policies

Historically, renewable energy in developing countries has come from direct donor assistance and grants for equipment purchases and demonstrations. In recent years a number of new approaches have emerged for promoting renewable energy in off-grid rural areas, including energy service concessions, private entrepreneurship, microcredit, and comparative line extension analysis.

Rural Electrification Policy and Energy Service Concessions

Many developing countries have explicit policies to extend electric networks to large shares of rural populations that remain unconnected to power grids (globally, an estimated 1.7 billion people). However, in many areas, full grid extension is too costly and unrealistic. Policies and rural electrification planning frameworks have recently started to emerge that designate certain geographic areas as targets for off-grid renewable energy development. These policies may also provide explicit government financial support for renewable energy in these areas. Such financial support is starting to be recognized as a competitive alternative to government subsidies for conventional grid extensions. Countries taking the lead with such policies include Argentina, China, India, Morocco, the Philippines, South Africa, and Sri Lanka.

One form this government support can take is so-called "energy service concessions." With a concession, the government selects one company to exclusively serve a specific geographic region, with an obligation to serve all customers who request service. The government also provides subsidies and regulates the fees and operations of the concession. Rural energy-service concessions may employ a mixture of energy sources to serve customers, including diesel generators, mini-hydro, photovoltaic, wind, and biomass. Argentina, Morocco, and South Africa have initiated policies to develop rural concessions, with ambitious targets of 200,000 rural households in South Africa and 60,000 in Argentina.

Rural Business Development and Microcredit

Private entrepreneurship is increasingly recognized as an important strategy to fulfill rural energy goals. Thus, rural electrification policies have begun to promote entrepreneurship. Promising approaches are emerging that support rural entrepreneurs with training, marketing, feasibility studies, business planning, management, financing, and connections to banks and community organizations. These approaches include "bundling" renewable energy with existing products. Bundling can reduce costs if vendors of existing products and services add renewable energy to their activities—and use their existing networks of sales outlets, dealers, and service personnel. Dealers of farm machinery, fertilizers, pumps, generators, batteries, kerosene, LPG, water, electronics, telecommunications, and other rural services can "bundle" renewable energy with these services.

In conjunction with entrepreneurship, consumer microcredit has emerged as an important tool for facilitating individual household purchases of renewable energy systems like solar home systems. Credit may be provided either by the system vendors themselves, by rural development banks, or by dedicated microcredit organizations. Notable examples of consumer microcredit for solar home systems have emerged in five developing countries. In Bangladesh, Grameen Shakti, a non-profit vendor, has offered consumer credit for terms up to 3 years. The Vietnam Women's Union offered similar credit terms for systems sold by private vendors. In Sri Lanka, Sarvodaya, a national microfinance organization, has offered credit on terms up to 5 years. In Zimbabwe, vendors sold several thousand systems on credit provided by the Agricultural Finance Corporation. And in India, new forms of rural microcredit have started to emerge. By 2002, the cumulative number of solar home system purchases made with credit in these countries had exceeded 50,000, but this was still a small fraction of the total number of solar home systems worldwide, estimated at 1.2 million.

Comparative Line Extension Analyses

Economic comparisons of line extension versus distributed renewable energy investment are also emerging in developed countries. At least four U.S. states have power line extension policies requiring that, in cases where utility customers must pay a portion of construction costs for utility power line extension to a remote location, the utility must provide information about on-site renewable energy technology options. Some of these policies require the utility to perform a cost/benefit analysis comparing line extension with off-grid renewable energy. Renewable energy options may be less expensive for rural customers, but without line extension policies, many customers would not be aware of this.

Public support for renewable energy expanded rapidly in the late 1990s and early 2000s. A wide variety of policies are designed explicitly to promote renewable energy, while other policies focus on power sector restructuring or environmental issues more broadly and have more indirectly affected renewable energy. Experience with renewable energy policies around the world is still emerging and more understanding is needed of the impacts of various policies. Thus, many policies could still be considered "experimental" in nature.

References

Bolinger, M., Wiser, R., Milford, L., Stoddard, M., and Porter, K. 2001. States emerge as clean energy investors: A review of state support for renewable energy. *The Electricity Journal* (Nov.), 82-95.

Dunn, S. 2000. *Micropower: The Next Electrical Era.* Worldwatch Institute, Washington, DC.

Hirsh, R.F., and Serchuk, A.H. 1999. Power switch: Will the restructured electric utility system help the environment? *Environment 41*, 4-9; 32-39.

International Energy Agency. 2003. *Renewable Energy Policies and Measures in IEA Countries.* Paris.

Martinot, E., Chaurey, A., Moreira, J., Lew, D., Wamukonya, N. 2002. Renewable energy markets in developing countries. *Annual Review of Energy and the Environment 27*: 309-348.

5

Energy Geopolitics

Since the industrial revolution the geopolitics of energy – who supplies it, and securing reliable access to those supplies – have been a driving factor in global prosperity and security. Over the coming decades, energy politics will determine the survival of the planet.

The political nature of energy, linked to the sources of supply and demand, comes to public attention at moments of crisis, particularly when unstable oil markets drive up prices and politicians hear constituent protests. But energy politics have become yet more complex. Transport systems, particularly in the United States, have become largely reliant on oil, so disruption of oil markets can bring a great power to a standstill.

Access to energy is critical to sustaining growth in China and India – not only to lift these countries out of poverty, but to keep pace with burgeoning populations. Failure to deliver on the hope of greater prosperity could unravel even authoritarian regimes, and even more so democratic ones, as populations become more educated and demanding. And it is these very factors that have turned the market power of energy suppliers into political power. Importers have come to compete for supplies, driving up prices, supplier wealth and the capacity to play roles in regional and international politics that go well beyond the GDP of countries such as Russia, Venezuela and Iran.

These traditional geopolitical considerations have become even more complex with global climate change. The United Nations' Intergovernmental

Panel on Climate Change has irrefutably documented that the use of fossil fuels is the principal cause of greenhouse gases that are driving up the temperature of the planet. Climate change will create severe flooding and droughts which will devastate many countries' food production, lead to the spread of various illnesses, and cause hundreds of thousands of deaths per year, particularly for those living in the developing world. Nearly two billion people were affected by climate related disasters in the 1990s and that rate may double in the next decade. At the very same time that countries are competing for energy, they must radically change how they use and conserve energy. The politics of that debate, particularly how to pay for the costs and dissemination of new technologies, and how to compensate those who contribute little to climate change but will most severely experience its tragedies, are emerging as a new focal point in the geopolitics.

Ironically, high oil and gas prices and the actions that must be taken to address climate change – namely, pricing carbon at a cost that will drive investment, new technology and conservation to control its emission – will drive another existential threat: the risk of nuclear proliferation. Higher energy and carbon prices will make nuclear power a more attractive option in national energy strategies, and the more reliant that countries become on nuclear power, the more they will want to control the fuel cycle. The risk of breakout from civilian power to weaponization would increase dramatically, as well as the risk of materials and technology getting into the hands of terrorists.

Confronting these challenges requires an understanding of the fragility of international oil and gas markets, but also of the nexus among energy security, climate change, and nuclear energy and proliferation.

Geopolitics of Oil and Gas

Political instability in and around countries considered marginal oil suppliers can cause major price spikes. That instability has created angst among oil importers and given even greater political power to oil suppliers. For example, when Turkey threatened on 17 October 2007 to take its fight against the terrorist organization, the PKK, into Kurdish Iraq, oil prices jumped from $87.40 per barrel to $94.53 per barrel by the end of the month. Yet Turkey is not an oil exporter, and Iraq produces only about 3 million barrels a day in a world market of 85 million barrels per day. Understanding the factors driving these fluctuations is at the heart of understanding the geopolitics of energy.

Oil consumption began to outstrip production in early 2006, which initially seems counterintuitive. Reducing oil inventories is favored when the future price of oil is predicted to fall. Conversely, inventories are built up when future prices are expected to rise. Consumption of these inventories accounts for consumption outstripping production. Alleviating this tight supply situation depends upon relatively unstable regions, while the security of transit through chokepoints creates vulnerabilities that factor into both security of energy supplies, but also environmental security. Nearly 25% of world oil exports pass through the Strait of Hormuz, nearly 15% through the Strait of Malacca, and nearly 5% through Bab el-Mandeb, the narrow strait connecting the Red Sea and the Gulf of Aden.

The most extreme political impacts of today's oil market realities are played out by Iran, Venezuela and Russia. Iran is developing a nuclear program despite UN Security Council Resolutions 1696, 1737 and 1747 demanding that Iran suspend the enrichment of uranium and fully disclose the nature of its nuclear program. When the International Atomic Energy Agency (IAEA) Board of Directors referred Iran to the UN Security Council (UNSC), as well as through the various UNSC Resolutions on Iran, countries from every part of the world have opposed Iran developing the capability to produce a nuclear weapon. Yet still Iran remains defiant. In part that may be out of the hope that Russia and China will block any serious sanctions, either because of their commercial interests in Iran, or because they generally resist setting a precedent for the UN to scrutinize individual national security decisions. But just as powerful is Iran's market power. Iran's oil revenues were $46 billion in 2005. They rose to $47 billion in 2006, and projections for 2007 and 2008 could put revenues in the $60 billion range.

Venezuela's influence must be seen in a wider context of globalization and its impact in Latin America. Globalization has helped millions in Latin America tap into technology, markets and capital in a way that has made many countries and people wealthier. But the gap between the "have's" and "have not's" has grown. Those who have not made it are also increasingly better educated – and resentful for what they don't have. That resentment is strongest among those making the transition out of poverty but who cannot see how to advance further. So they become vulnerable to populism. When given a chance to vote, many will use their ballots to express their frustration. It is in this context that Venezuela and Hugo Chavez have brought their wealth to bear. Within Brazil and Mexico, Chavez's message of populism and his support for local leaders have the potential to galvanize local frustrations. In Bolivia and Nicaragua, the Chavez myth seen from the outside seems to suggest that the poor could be given more at little cost.

Not every Latin American country has gone down Chavez's populist route, but he is presenting new challenges to a regional order based on democracy and market principles. For democrats in the region, the first challenge is to ensure that there is not a backlash against democracy from those leaders and countries that feel threatened by popular frustration. The second is to reform governance and policies to give the "have not's" a sense that they can have a better future. Whether Latin American leaders can educate their people to create the capacity to benefit from globalization, whether governments can target subsidies to those who need to be pulled into society, and whether the United States will open its markets to technologies, services and products – these factors together will fundamentally affect perceptions of democratization in the region, and whether it is a source of stability or a vent for populism.

Russian Scenario

Given Russia's veto power in the United Nations Security Council, its unique supply position for gas to Europe, and its control over one of the two largest nuclear arsenals in the world, it is important to understand how energy has transformed Russia internally and its role in the international community. In addition to being the world's second largest exporter of oil, Russia controls over a quarter of the world's proven gas reserves at 1,680,000 billion cubic feet. Europe imports 23% of its gas from Russia.

Diversification of transport routes is costly and takes time. For example, the North Europe Gas Pipeline (NEGP) will connect Vyborg, Russia, to Griefswald, Germany, and consist of two parallel pipelines, the first to be commissioned in 2010 and the second in 2012, with a total capacity of 55 bcm/year. The NEGP is projected to meet nearly 25% of Europe's additional gas import needs by 2015, but will cost an estimated $5 billion. And, of course, this specific project will only further entrench Germany's dependence on Russian gas.

Russia currently ranks 8th in the world in terms of proven petroleum reserves, at 60 billion barrels. It ranks second only to Saudi Arabia in terms of oil production, at 9.4 billion barrels per day. Russia's $147.6 billion Stabilisation Fund, in which revenues from export duties on oil and taxes on oil mining operations accumulate when the price for Urals oil exceeds the set cut-off price – which is intentionally kept relatively low in order to ensure that the bulk of oil-generated revenues accrue to the Fund – was established in 2004 as a means of hedging against "Dutch Disease" and paying off debt. The 2007 and 2008 budgets were based on a cut-off price of $27/bbl, although this price may be revised in light of rising oil prices. Russia has also accumulated $425 billion in hard currency and gold reserves.

This energy wealth enabled Russia to pay off $50.7 billion in debt ($3.33 bn for early debt repayment to the IMF, $43.1 bn for debt repayment to the countries-members of the Paris Club, and $4.3 bn paid to Vensheconomobank (VEB) for loans provided to the Ministry of Finance in 1998-1999 for servicing state foreign debt of the Russian Federation). Corporate debt, totalling $384.8 billion at the end of the first half of 2007, has risen 24% since the beginning of the year. The recent debt growth rate is more of a concern than the actual amount of debt, given that Maastricht standards cite corporate debt over 30% to be dangerous for the macroeconomics of a state. Moreover, Russia has refused to ratify the Energy Charter Treaty, which would guarantee transit rights for energy through Russia regardless of the owner and preclude cutting off energy supplies as a political weapon.

Its energy market power has allowed Russia to consolidate political power internally and has made Russia immune to normal external checks on the exercise of power. Within Russia, Putin has been able to control the appointments of governors and the upper house of parliament. He has orchestrated a change in rules for parties to get into the lower house of parliament, in turn tightening the ties between political parties and the Kremlin. He has appointed individuals linked to the Kremlin to corporate leadership positions in, among others, the gas, oil, rail, airline, shipping, diamond, nuclear fuel and telecommunications industries. The Kremlin has also consolidated control over most broadcast media, and has been able to get the courts to do its bidding on cases against power rivals such as Mikhail Khodorkovsky.

Externally, Putin expertly managed a bidding war for his attention between President Bush and Presidents Chirac and Schroeder when both the U.S. and "old Europe" sought to get Putin on its side in the war on terror and the Iraq War. By the time that the United States and Europe began focusing on Putin's consolidation of Russian politics in late 2004, new political realities had been created, backed by a newfound stability in oil wealth that would only grow and become reinforced by Europe's dependence on Russian gas. Today, the result is that Putin ignores international entreaties over the Kremlin's control of domestic politics, Putin presents himself as the protector of international law and order against American aggression, Russia uses energy as leverage in its negotiations with what it sees as upstart neighbors (Ukraine, Georgia), and it continues to resist Europe's entreaties for comparable rights for its investors in Russia.

But arguably the most complex and significant evolution of Russian energy power combined with its political weight on the UN Security Council has been its role on Iran. On the one hand, Russia has stated that it has no interest in Iran acquiring nuclear weapons, and it has been part of the political negotiating group consisting of the five permanent Security Council members and Germany. Yet Russia has resisted the imposition of tough sanctions against Iran, seeking to carve out exceptions for Russia's sale of civilian nuclear technology for Iran's Bushehr nuclear power plant and to weaken UN sanctions against Iran, providing cover for China to follow suit. Recently, Russian officials or former officials have indicated that they see prospects for the International Atomic Energy Agency to close out the file concerning the historical questions about Iran's nuclear program and, according to these individuals, that would require returning the Iran case from the UNSC to the IAEA.

Russia, in effect, has positioned itself to either unravel or make viable an effective diplomatic package against Iran. If it splits the "P5 plus 1" by insisting that the UNSC has no role to consider sanctions against Iran, Russia will almost surely split any effective diplomatic effort, give Iran further leeway, virtually ensure that Iran develops nuclear weapons capability, and raise the risk of a U.S., Israeli or other military action against Iran. Yet Russia also has the capacity to be part of a package that could make clear to Iran – and just as important to the Muslim world – that the international community is not blocking Iran from a civilian nuclear program. To the contrary, Russia's cooperation could make possible the offer of a more advanced civilian nuclear plant, nuclear fuel and the reprocessing of spent fuel, ideally in the context of a package that could be extended as well to other states seeking civilian nuclear programs.

The Iran case, and Russia's role in it, bring together most key elements of today's complex geopolitics of energy: market power to act in isolation, multiplying energy power by seeking to bloc multilateral instruments, the emerging risks associated with civilian nuclear power, and limited short-term recourse to exercise rule-based order to control energy market power. For energy consumers – and those who see the wider risks of vesting so much political sway in energy-rich states – the short-term options are limited. Better management of reserves could help, and bringing China and India into a reserve management system would seem crucial since they are biggest drivers of increased oil demand, yet they are out of the International Energy Association's reserve management system. The more critical changes are in the medium term, through conservation, alternative fuels, massive lifestyle changes, new building codes, and new technologies that burn less energy. It is these very types of policies that are also central to a different yet even more existential aspect of the geopolitics of energy: climate change.

Geopolitics of Climate Change

Avoiding the destruction of the planet through the emission of greenhouse gases is one of the most complex challenges that we, as the collective human race, have ever created for ourselves. Our very survival is at stake. The difficulties lie in the intersection of earth sciences, technology, economics and politics. The emission of greenhouse gases will have the same impact regardless of the source – whether Beijing, Detroit or Newcastle. Hence it is impossible to solve the global problem without involving all states. The problem of climate change, due to the concentration of greenhouse gases in the atmosphere (CO2-equivalent), was created by the industrialized world, so emerging market economies resent that they must share the cost in addressing the problem. Yet emerging economies are the fastest growing source of greenhouse gas emissions. Worse yet, the biggest catastrophic impacts will be on developing countries such as Mali and Bangladesh that are not driving the problem in any way.

The Intergovernmental Panel on Climate Change (IPCC) established that maximum temperature increase that the world can sustain without causing irreparable damage is about 2.5 degrees centigrade by 2050. There is less certainty about what concentration of CO_{2e} will avoid going over a 2.5 degree temperature increase, but the estimates generally fall in the range of 450-550 parts per million (ppm) of CO_{2e}. The lower the level, the costlier and harder it is to achieve. We are currently at a level of about 420 ppm of CO_{2e}. There is also uncertainty about the level of annual reductions in greenhouse gas emissions that are needed in order to stabilize the atmosphere at a concentration of 450-550 ppm of CO_{2e}, but estimates cover a range from 50-85% in annual reductions of CO_{2e} emissions relative to 1990 levels.

The objective of a climate change policy must be to create the incentives that will drive changes in technology, technology dissemination, consumption patterns, and new developments in how energy is produced in order to reduce the annual emission of carbon so that the atmosphere does not exceed a concentration of more than 450-550 ppm by 2050. That is a monumental task. If one were to assume the continuation of current practices and technology, estimates indicate that greenhouse gas emissions could *increase* by 25-90% by 2030, much less decrease on the order of 50% or more annually by 2050.

We currently do not have the technologies and policies to achieve this target. Conservation, efficiency, alternative fuels, and cleaner use of fuels all have to be part of the equation. But the combinations we currently have available do not achieve the desired end point. In order to succeed, the international community must find a way to price carbon in order to curb consumption, spur technological innovation, affect fuel choices, and stimulate investment. Some argue that, in the long term, there must be a stable long-term price for carbon of at least $30/ton to achieve the necessary economic and technological incentives.

Yet pricing carbon has been another source of geopolitical divide. No country has adopted an explicit tax on carbon on the scale of $30/MT. Cap and trade systems in Europe or those emerging in regions of the United States do not yet come close to this level of implicit carbon price. Within the U.S., our more progressive states have adopted standards for the use of renewable fuels and efficiency. Some states like Florida and California have set targets for overall GHG emissions. They are creating an implicit cost for carbon, but they are not setting the stable and explicit price signals that are needed for innovation. Japan, for example, has called for a 50% annual reduction in CO_2 emissions by 2050, but the Japanese Government has kept a cap and trade system and a carbon tax off the table as policy options.

From the debates over policy, economics, technology and science have emerged four geopolitical blocs on climate change, and perhaps a fifth waiting in the wings. The first is anchored by Europe and, with less fervor, Japan, and supports the adoption of binding emissions targets. The second is driven by the United States and supports setting a long-term goal and nationally binding medium-term commitments, but not an internationally binding treaty that holds countries collectively to account. The third consists of the emerging market economies led by China and India, it has resisted any form of binding international targets, and it has focused its demands on technology dissemination and financing for the cost differential for clean technologies. The fourth group is that of developing countries bearing the brunt of flooding, desertification and other catastrophic effects of climate change, and their demands focus on financing to adapt to the impacts of climate change. Perhaps the emerging fifth group may be that of energy suppliers who see the world shifting away from the use of fossil fuels. They could emerge either as facilitators of transition if they invest their wealth in technology dissemination and thereby positions themselves as winners in a greener international environment, or they could be spoilers who seek to drive up prices and profits to capture the greatest earnings in the course of transition.

Among these groups, the United States has the capacity to be a pivotal figure. China and India will not move toward more responsible international policies if the United States does not set the example. Along with Europe and Japan, the United States has the capacity to demonstrate that green technology and conservation can be compatible with growth and a foreign policy that is more independent of energy suppliers. The United States also stands to benefit from accelerated commercialization of green technologies and the development of global markets in energy efficient and clean energy technologies. But the ability of the United States to lead will fundamentally depend on domestic action: whether it will undertake on a national basis a systematic strategy to price carbon and curb emissions. If it does, the scale and importance of the American market can be a driver for global change. If not, then the United States will find that over time the opportunity for leadership to curb climate change will be replaced by crisis management as localized wars, migration, poverty and humanitarian catastrophes increasingly absorb our international attention and resources, eventually coming back to our own borders in a way that will make the Katrina disaster seem relatively small.

Geopolitics of Nuclear Proliferation

Perhaps the most serious existential risk that parallels that of climate change is that of nuclear technology and materials getting into the hands of rogue states or terrorist organizations. That could result in the devastation of cities or nations and set off reciprocal actions that lead to levels of destruction that were only foreshadowed at Hiroshima and Nagasaki in World War II. High fossil fuel prices, the risks associated with energy suppliers and transport routes and, ironically, policies to combat climate change – namely, the pricing of carbon – could accelerate a drive for civilian nuclear power that could increase that existential risk. For economic, environmental and security reasons, we should expect more and more countries to incorporate nuclear power into the mix of their power generation capabilities.

Today just twelve countries, out of the sixty states with some form of nuclear capacity, can enrich and commercially produce uranium. Arguably, nine countries currently have nuclear weapons. Imagine if the number producing enriched uranium were to double or triple if developing nations sought to enhance their energy security through a misguided sense of energy self-reliance while adopting carbon-free nuclear technology to produce electricity.

That calls for an intensified effort now, before it is a crisis, to strengthen the firewalls between civilian nuclear power and weaponization programs.

The goal must be to give aspirants for civilian nuclear power confidence to obtain nuclear fuel through an international fuel bank and to forego enrichment programs, while placing their entire nuclear programs under the International Atomic Energy Agency (IAEA) Additional Protocol. Such measures may not stop Iran's nuclear ambitions, but they may help other countries from breaking out from civilian nuclear programs to weaponization. They will also reduce the risk of nuclear material leaking into the hands of rogue states and terrorists. And to achieve the credibility to lead the international community in forging such a revitalized regime against proliferation, the United States will need to follow through on the promises it has made to what the non-nuclear weapons states see as "horizontal proliferation", namely ratification of the Comprehensive Test Ban Treaty (CTBT).

Realizing a safer international nuclear regime will require revitalizing the bargain between nuclear and non-nuclear weapons states under the Nuclear Nonproliferation Treaty (NPT). Article 4 of the NPT assures non-nuclear weapons states the right to peaceful civilian applications of nuclear power if they adhere to the treaty's provisions and forego the pursuit of nuclear weapons.

Since the drafting of the NPT in 1968, experience has demonstrated ways in which monitoring and surveillance should be enhanced to reduce the risk of leakage, and these measures have been incorporated into a voluntary Additional Protocol. In return, nuclear weapons states committed under the NPT to reduce their arsenals and seek eventual nuclear disarmament.

It is the disarmament part of this agenda that Secretaries Kissinger, Schultz and Perry, along with Senator Nunn, have proposed in their renewed call for the elimination of nuclear weapons. Even for those who think that full nuclear disarmament is unworkable or unwise, U.S. ratification of the CTBT is the most critical step to restore the credibility and vitality of the bargain the NPT established between vertical (across states) and horizontal (deepening within nuclear states) proliferation. At the 1995 NPT review conference, non-nuclear weapons states accepted an American commitment to the ratification of the CTBT as a basis for the indefinite extension of the NPT – in effect, a deal for their permanent commitment to forego nuclear weapons. In order to advance now the actions needed to curtail the "vertical proliferation" of nuclear weapons, the United States cannot ignore its 1995 commitment on CTBT.

A new package is needed on proliferation and testing that includes:

- A commitment by NPT signatories to accept the Additional Protocol,
- The development of an international fuel bank under the IAEA that would assure nations supply to nuclear fuel as long as they observe the NPT's provisions,
- A means to centralize the control and storage of spent nuclear fuel, and
- A ban on testing that would complicate the ability of any aspirant for nuclear weapons to break out of a civilian nuclear program.

The ban on testing is pivotal in the geopolitics of nuclear power. A comprehensive test ban would have the greatest impact on states that want to use civilian programs as a platform for the development of nuclear weapons. Nuclear weapons states have other means to service and replenish their arsenals. Those truly committed to civilian nuclear power should not have a need to enrich, and in most cases the scale would be sufficiently small that it would not make economic sense for them to do so. If any entity were to test a nuclear weapon, it should be immediately detectable, and it should trigger sharp multilateral pressure to abandon the program. This was the case with North Korea, where China, the United States and Japan quickly secured UN condemnation and sanctions after North Korea's nuclear test in October 2006.

A comprehensive test ban creates the incentive to sustain the status quo among nuclear states, and to constrain states from developing nuclear weapons capacity. The CTBT isolates those who seek to advance their ambitions for nuclear weapons. Russia would need to be part of this package – as a supplier of fuel, and a secure source for storage and reprocessing – with massive commercial benefits to Russia. The United States should seize on this opportunity – if not now, then under a new President in 2009 – to ratify and implement the CTBT, and in so doing strengthen American leverage to broker an international package to stop nuclear leakage and curtail the risk of breakout from civilian programs.

Challenges and Future Trends

For more than a century, energy, politics and power have been clearly intertwined as a force in international security. The stakes are only getting bigger as the issues go beyond national prosperity and security, to the viability of the planet. Policy makers and citizens must understand the nature of this change and recognize that inaction – simply not attempting to forge coalitions and to guide constructively how states use energy – will be catastrophic.

It will be crucial to resist allowing short-term electoral cycles in the United States or elsewhere to drive energy policy and politics. Inevitably, some politicians will make unrealistic and unattainable calls for energy independence. That is simply not possible in an interconnected world that requires access to global markets, capital and technology, whether a nation is a net importer or exporter of energy.

In the short-term, diplomacy and effective reserve management will be critical tools that are not fully developed. Expansion of the IEA's reserve management system to China and India, and technical support to help them coordinate with others, will be an important confidence building measure for states that, at present, see themselves pitted against the rest of the international community. Energy diplomacy also needs to be made a central foreign policy consideration. Key questions include:

- Where can nations jointly benefit from further exploration and development?
- What transit systems merit international cooperation and investment?
- Are there regional security arrangements that can mitigate risk and create shared incentives across states, especially in the Middle East, the Gulf and Central Asia?
- Can the P5 reach an understanding to suspend the use of their veto rights on issues related to energy politics in order to stimulate a full debate around tough questions that get sidetracked through veto threats?
- Should nations commit to an E15 format, composed of the largest economies and energy users, as a means to force a focus and sustained agenda on the policies and politics behind energy supply and use?
- How do domestic energy and economic growth concerns drive the foreign policy choices of China and India and their roles in multilateral institutions?

Focused answers around these questions could be the foundation for national, regional and international energy strategies that foster cooperation around energy issues, rather than allow short-term political considerations shape what generally may appear to be zero-sum competitive outcomes.

In the medium and long-term, both geopolitical interests and environmental sustainability call for a radical departure from current patterns in the use of fossil fuels which, for most states, compromise national security, and for all nations threaten the planet. A shared medium-term strategy among states to foster convergence around political, environmental, energy and economic goals should include:

- Measures to price carbon emissions and to coordinate prices across states, if not create transnational carbon markets.
- Financing and policy measures (e.g. addressing liabilities associated with carbon capture and sequestration) to support the development, testing, demonstration, commercialization, and dissemination of clean and efficient technologies that can transform the terms of debate on energy use and climate change.
- Means to stimulate investment in clean technologies: to reduce private sector and temporal risk for the developed countries, to finance the differential between clean and traditional technologies for emerging

economies, and to develop infrastructure and adapt to climatic changes in developing countries.

- Common international standards for firms to disclose the use of carbon and establish guidelines for emissions per unit value of output in order to promote public accountability and guide investment decisions.
- A new form of international framework for climate change that reflects the complexity of the interaction of technology, economics and politics and leads to better and tighter standards for performance over time.

On the nuclear side, no issue is more important than creating a strong firewall now between civilian power and weaponization programs, before more countries seek to break out from civilian programs. Hard as this may be, it will be easier than getting new entrants into the ranks of nuclear weapons states to disarm. For this process to start, the United States must start with the ratification of the Comprehensive Test Ban Treaty, with India and Pakistan acting in concert with the United States.

These are major challenges. They are not unattainable. If such actions are taken now, we stand a chance to get the geopolitics of energy to move the international community toward constructive long-term outcomes. If not, we will allow the geopolitics of energy to make all nations less secure, and bring into question the very viability of our future.

References

Friedman, Thomas. L. 2006. 'The First Law of PetroPolitics' *Foreign Policy*, May/June 2006.

Gilpin, Robert. 1987. *The Political Economy of International Relations.* Princeton: Princeton University Press

Hinnebusch, Ray. 2003. *The International Politics of the Middle East.* Manchester: Manchester University Press.

Jafar, Majid. 2004. 'Kazakhstan: Oil, Politics and the New 'Great Game'. In Shirin Akiner (ed.), *The Caspian: Politics, Energy and Security*. London: RoutledgeCurzon.

Karasac, Hasene. 2002. 'Actors of the New 'Great Game.' Caspian Oil Politics.' *Journal of Southern Europe and the Balkans*, 4 (1).

6

International Competition for Energy Resources

The end of the Cold War can be argued as bringing an end to the 'old wars', and as international relations entered into the 21st Century, it was argued that there are 'new wars' in this new age. The time of resource wars has begun and the further we advance in time, the more often we will sec conflicts concerning vital resources like oil and gas. It is inevitable that there will be a lack of resources available for every nation-state, unless there is a discovery of an alternative plentiful resource. The emergence of the oil boom in the 60s-70s, saw state competition defining the politics of energy. Nevertheless, over time markets have been established to control the price of oil, and improve the producer-consumer relations; developments which in turn ensure that markets trading oil on a global scale have become more transparent. This paper will critically engage with the realist claim that state competition still defines the global energy of politics rather than markets, as claimed by liberalism.

The typical proclamations made by realist are that the international realm is "characterised by anarchy, distrust, and the ever-present prospect of war" . It is evident throughout history that states are led by self-interests, and national interest does play a very bias role when states make decisions in the international arena. Kenneth Waltz, the father and founder of neo-realism, contended that the international system is anarchical, also claiming that the structure of the system is focused on the balance of power between states. The custom of realpolitik "prioritises the interests of the sovereign", and the "key goal of statesmen seeking to preserve international stability is to contain the ineluctable drive for power by states".

The openness and control of energy is a major element of national interests and state competition for energy resources. This is due to energy resources becoming scarce because of the arguments of 'peak oil' theory, which only encourages states to compete for access to these resources. "Chinese leadership see that the insecurity of the Malacca straits, and the prospect of a military embargo of its oil supplies, represents a fundamental threat to China's core national interests". We are already witness to national interests and eagerness of a state for fuel with the resurfacing for the 'scramble for Africa' mainly from China, which sees Africa as very opportune for its national and strategic interests.

The U.S. dependence of oil imports can be seen as the "'Achilles heel' of American power". However, the US was not always dependant on its oil imports: "until the end of the Second World War, the United States was by far the largest oil producer in the world and self-sufficient in oil supplies". This drastically changed however, as state interests and greed lead the U.S. to comprehend that they cannot always be sustainable on the production of their own oil, and as a result they became one of the first countries to start exploration of oil in the Middle East and become an importer of oil, allowing them to keep reserves of their own oil. "Control of oil may be seen as the centre of gravity of US economic hegemony". The US most definitely has a "degree of influence" when it comes to oil explorations, especially in the Middle East. This was proven after 1945 when the U.S. strategically took the steps to be the first nation to start exploration of oil in the Middle East, and since the U.S. was the first state to do this, "Saudi Arabia and the Gulf states, in effect, exchanged military security for cooperation on pricing and production decisions ..., thereby reconciling their economic interests with those of the oil majors and the major consuming countries."

This meant that the U.S. for a lengthy period had negotiated long term contracts with state owned oil companies like Saudi Aramco for an incredibly low price. The U.S. has such a degree of influence in the oil markets that even in the OPEC cartel it has an observer status. "Since the OPEC nationalizations of the 1970s, US policy towards the international oil industry outside America rested on a combination of support for its companies as the producers and distributors of traded oil and cooperative relations with swing producers-especially Saudi Arabia- to stabilize the international markets".

Over the past 20 years the Saudi's have published official state figures of their oil reserves and they have steadily been similar, this fabrication of statistics has occurred due to the U.S.-Saudi interests in dictating the global oil markets for their benefits and needs, enabling the two nations to control the global price of oil. If actual statistical figures would show that the Saudi oil reserves have dwindled since American exploration then there would be significant oil price hikes. This was recently proven when the political uprising occurred in Libya against the dictatorship of Col. Gaddafi and there was a fear that it could cause an oil price hike due to shortages of oil supply as a result of the sanctions and embargos and instability in the region. However as always the Saudis came to the rescue and increased output to balance the market price of oil and to continue the flow of supply of oil. The same can be said with the current threats of Iran who claim that they will block the Strait of Hormuz; of course the Saudis will come to the rescue again.

State competition plays a vital and influential role in the global politics of energy. This is due to the basic history of statism, and throughout history we see the greed and self-interests of states and how the great super powers always fight to regain control of their power or to reaffirm their position on the international stage. U.S. hegemony is critically based around importing oil and the preservation of its own oil supplies, as the U.S. has realised that to sustain its power as the world's number one it has to be able to have a plentiful supply of resources. If the global politics of energy was more market orientated then why was the U.S. so adamant during the Cold War when the Soviets invaded Afghanistan, to go to the Middle East and secure the oil fields? This was due to the fear of the U.S. losing out on the oil from the Middle East to the great rival and power bloc towards the U.S., the Soviet Union. As we move forward today, the global politics of energy has intensified in the state competition with the rising super that is China, which not only is rivalled by the U.S. but faces strong opposition in the name of India.

International Relations Theories

IR theories seek to identify key regularities and patterns of interaction in the realm of international politics and to provide parsimonious models to explain the nature and underlying structures of that interaction. These theories generally have, whether implicitly or explicitly, a normative dimension – they say something about how international politics should be conducted and what the world should look like. IR theories have traditionally been seen as competing and incommensurate, as representing alternative 'paradigms', and their historical evolution often involves critiquing the perceived flaws of alternative theories. The traditional division of IR theories, which was common in the Cold War period, was between realism, liberalism and Marxism/structuralism. Since the end of the Cold War a dominant new theoretical approach is that of social constructivism, which emerged as a critique of both realism and liberalism (or more precisely neo-realism and neo-liberalism) and filled a gap with the decline of the intellectual appeal of Marxism. However, this radical dimension has been increasingly filled by a variety of theories – critical theory, feminist IR theory, historical sociology, post-structuralism. The overall picture is currently quite complex with an array of differing methodological, explanatory and normative theoretical approaches.

Despite this profusion of theoretical approaches and the clear and evident importance of energy in international relations, it is striking that there has been limited direct application of IR theories to understanding energy- and mineral-related conflicts and modes of collaboration and competition. As Brenda Shaffer points out, the principal journal in international relations and security studies, *International Security*, has only published eight articles devoted to energy in its 30 year history. Admittedly, journals like *Foreign Affairs, Foreign Policy*, *International Affairs* and *Washington Quarterly* do often have energy-related articles but these journals are primarily policy-related rather than theoretical journals. Nevertheless, most of the IR literature on energy issues, particularly as related to explaining the potential for conflict and cooperation, is *implicitly* theoretical, with the main arguments and policy prescriptions underpinned by certain fundamental theoretical assumptions.

Realism, neo-Realism and Geopolitics

Realism is often seen historically as the dominant IR theory and this is certainly correct in terms of the study of security, conflict and war. This is reflected in the fact that *International Security*, as noted above the flagship IR journal, is dominated by realist and neo-realist authors. Classical realism includes the key early and mid-twentieth century scholars who developed a notion of the 'tragic' nature of international politics, arguing that there was a radical difference between politics within a state and politics between states since inter-state politics lacks any overarching sovereign arbiter who is able authoritatively to repress the inexorable drive for power and the natural human tendency towards aggression. The logical consequence is that the international realm is chracterised by anarchy, distrust and the ever-present prospect of war. Much of realism's initial momentum and subsequent popularity came from its critique of inter-war liberalism (or so-called idealism) and the optimism expressed by may liberals that international relations could be transformed through developing international law and international institutions such as the League of Nations. In 1979, Kenneth Waltz provided a more rigorous and parsimonious model of realism, known as neo-realism, whose main assumptions were that the international system is anarchical, that the structure of the system is determined by the distribution of power between states (the balance of power), and that the internal nature of the state has no material structural impact on international relations.

Realism's theoretical principles draw from deeper historical traditions of thinking about international politics and these help to explain the theory's popularity and theoretical dominance. This includes the tradition of *realpolitik* developed from Machiavelli onwards, which prioritises the interests of the sovereign, and where the key goal of statesmen seeking to preserve international stability is to contain the ineluctable drive for power by states, and the conflicts this inevitably produces, through the preservation of a durable balance of power.

As Kissinger has described, this was the foundation of the European order in the 18th and 19th century. It was an approach to international politics he also sought to resurrect to develop his own foreign policy principles when he was a highly influential US Secretary of State in the 1970s. Another tradition which realism draws from is that of geopolitics which includes the work of people like Mahan , Mackinder , Haushofer, Harold and Margaret Spout , and Lipschutz. This tradition draws from geography as well as IR and strategic studies and highlights the spatial dimensions of state power and identifies a continued international struggle for influence and control of critical geographical and geopolitical spaces, whether that be the Eurasian 'heartland' favoured by Mackinder or the international sea lanes promoted by Mahan.

Much of the literature on the politics of international energy adopts implicitly a realist and geopolitical theoretical approach, even if this is rarely explicitly developed. The key underlying assumptions and arguments of those who adopt this approach can be reduced to the following:

- Access to and control of natural resources, of which energy is the most critical, is a key ingredient of national power and national interest
- Energy resources are becoming scarcer and more insecure (drawing often from the 'peak oil' thesis and the 'resource curse' and 'resource wars' literature)
- States will increasingly compete for access and control over these resources
- Conflict and war over these resources are increasingly likely, if not inevitable.

A good illustration of this general approach can be seen in the work of Michael Klare who has written prolifically on the international politics of energy and is probably the best-known and most popular writer in the field of IR and energy. The core arguments of his various books are essentially realist and can be distilled to:

- In the post-Cold War period, with the end of the ideological clash between socialism and capitalism and the rise of new economic powers, international relations is increasingly focused on gaining or maintaining access to and control of valuable natural resources, which is inextricably linked to the post-Cold War shifts in the balance of power. This is a major source of conflict between the most powerful states: US, China, Russia, EU, Japan, India…etc
- Natural resources, most notably oil, is becoming increasingly scarce due to rising demand in Asia and the prospect of 'peak oil'.
- Much of the world's supply of oil, and much of its new supplies such as in Central Asia and Africa, are located in weak, fragile states with multiple inter-state disputes and conflicts and where political and religious extremism is rising. Oil wealth has the paradoxical effect of making these states more powerful international actors, due to their control of vital resources, but also more dysfunctional, more 'dissatisfied', revisionist, authoritarian and anti-Western. A link is to be found between resource wealth and the post 9/11 growth of radical Islam and the threat of international terrorism.

- International conflict over oil and other natural resources is thus becoming more and more likely.

This general overarching thesis is undoubtedly a powerful and persuasive framework which captures the political imagination of many analysts and policy-makers, and which needs to be taken into account even by those who might disagree with the underlying assumptions. Such an approach feeds, for example, the concerns of the Chinese leadership see that the insecurity of the Malacca straits, and the prospect of a military embargo of its oil supplies, represents a fundamental threat to China's core national interests; similarly, it underlay the concerns of the US Congress that CNOOC's bid for UNOCAL in 2005 would, if successful, represent a critical threat to US national interests and its energy security. It is a theoretical frame which suffuses military planning, such as that of the Pentagon or the PLA or the Russian armed forces, and promotes national defence strategies which incorporate policies to defend perceived vulnerable energy supply sources and transportation routes. It also feeds into more alarmist policy and journalistic accounts of international relations where there has been a burgeoning literature about the new 'Great Game' in Central Asia, which pits Russia, China and the West in a zero-sum game for control over the region's energy resources. Similarly, the emergence of a renewed 'scramble for Africa' which focuses on the increased global interest in the natural resources of Africa, most notable of which is oil, and which has made this region regain strategic importance and which has incited great power competition. This realist-driven energy conflict approach also suffuses Western concerns over the rise of China, the fears of Chinese expansion in Central Asia, Africa, Latin America, and the prospect of increased conflict between China and its regional neighbours, Russia, Japan and India.

Despite the evident power and attractiveness of such realist-driven analyses of international energy politics, critics have identified significant shortcomings and weaknesses in these accounts. The most critical of these include:

- The over-emphasis on the strictly military dimensions of power. Some of these criticisms have emerged from within the realist tradition. For example, the neo-realist Robert Gilpin has been highly influential in highlighting the economic factors which are critical to national power and how, for example, multinational corporations are key components of a country's overall national power. Susan Strange can also be considered to be broadly realist in that she emphasises the primacy of states in the international system (even if firms also play a major role) but argues that the power of states is driven by four dimensions of power - military, production, finance and ideas – and that any country's international standing and relative power must be assessed across these four dimensions. The role of 'ideas' in global politics has been further popularised by Joseph Nye and his argument that there are 'soft' as well as 'hard' dimensions of power, and that the attractiveness of national culture and ideology are critical facets of national power. Nationally-related ideological constructs, such as the 'Washington Consensus' (neo-liberalism), the 'Beijing Consensus' (authoritarian state capitalism) or the 'European social model' (social democratic capitalism) are examples of this soft power competition.

- A linked area of criticism is that realism tends to be too state-centric. This is area where most of the alternative theories converge in critiquing realism. In terms of international energy politics, this involves a criticism that too much attention is accorded to states and inter-state competition and too little attention to the autonomous role of transnational actors (such as the transnational oil and mining companies) and of local actors (local and sub-national communities affected by or seeking to gain control of mining activities). These critics argue that it is this more complex and nuanced interaction between the transnational/national/local which is often left out of realist-inspired accounts.
- A further common criticism is that realism tends to be overly deterministic. The assumption that resources are scarce and that inter-state conflict is inevitable are generally taken as given in these accounts. The widespread challenges to the 'peak oil' thesis, the role that technological innovation plays in relieving scarcity, the potential for substitutability – these are dimensions which are rarely articulated in realist analyses. The role of international markets and regional and international institutions in managing and diffusing conflicts are discounted. The prospect that countries might seek to avoid war, and are not driven inexorably towards conflict due either to natural aggression or the inexorable logic of the 'security dilemma', are also similarly often ignored. This is, for example, a core argument of the social constructivist approach which sees that there are varying and

differing ways in which states and international actors can decide to interact and that they do not necessarily have to operate within the rigid realist constraints of anarchy; to use the famous title of an article by Alexander Wendt, 'Anarchy is what states make of it' and there can be more cooperative alternatives, such as has developed in the context of the European Union.

The Liberal Tradition

The liberal tradition in International Relations can be usefully viewed as a conscious critique of the realist approach to international politics and its allied traditions of *realpolitik* and geopolitics. Liberal internationalism developed through its explicitly anti-realist explanations for the causes of World War 1 – that the tragedy and slaughter of WW1 was precisely due to the fact that states blindly pursued realist policies, such as secret diplomacy, the obsession with the balance of power, the ignoring of public opinion, and the failure to develop international legal norms and institutions. At its core, the liberal tradition in IR rejects the realist assumption of a radical disjuncture between domestic and international politics and that different moral and practical principles apply in these two realms. Fundamental liberal principles, most notably the obligation to respect individual autonomy (human rights) and that political institutions should be developed institutionally to respect those rights (democracies), are seen to apply equally to international as well as domestic politics. A key argument of liberal IR is that democracies conduct foreign policy differently than authoritarian regimes and that democracies most notably do not fight wars against one another, the so-called 'democratic peace' thesis. Liberals reject, in this regard, the deterministic and pessimistic realist view of international politics (as being a form of 'tragic' politics) and argue that progress can and should be made; universal human rights can be promoted and defended; countries can shift from being authoritarian to democratic states; and that it does make a difference if international politics is conducted between democracies; that regional and international institutions and regimes can be developed which are conducive to increased cooperation; that global prosperity can be attained if markets are left open and trade is liberalised; and that anarchy can be overcome and war avoided.

As noted above, some of the historic drivers of realism were themselves a counter-reaction to the perceived moralism and excessive optimism (idealism) of liberal exponents of international politics. There is thus a continual dialogue and debate between realists and liberals with, at times, some shifts in position to accommodate their respective critiques. Such a partial compromise on the liberal IR tradition can be seen in the 'neo-liberal' or liberal institutionalist approach which emerged in the 1970s and 1980s and which accepted the realist assumption of anarchy but nevertheless argued that regimes and institutions, based on liberal principles of transparency and legally binding norms, could lead formerly antagonistic actors to adopt cooperative behaviour and promote positive-sum results. Regime theory was the main conduit for this institutionalist turn and which sought to articulate more precisely the transmission mechanisms whereby international economic cooperation could stimulate political cooperation the energy field. The European Union became the main liberal institutionalist paradigm of the ways in which a regional institution could overcome state sovereignty through economic and political interdependence (the 'spillover' or functionalist thesis) and thus reverse the realist-driven propensity for inter-state war and make war 'inconceivable' in Europe.

In terms of international energy politics, the liberal approach can be implicitly or explicitly be seen to inform two major sets of work; the first could be categorised as the liberal quest to expose the '*dark underbelly*' of the international energy industry; the second as the liberal policy prescriptions of '*what needs to be done*' so as to generate a more open and cooperative set of arrangements in the international management of the international energy industry. As noted above, liberal approaches tend not to make a hard and fast division between international and domestic politics or between economics and politics, so there tends to be much overlap or blurring of the borders between liberal IR and comparative politics and between economics and politics/IR.

The 'Dark Underbelly' involves an exposition, from a predominantly liberal perspective, of a number of the perceived embedded illiberal practices and perversions of the politics, economics and international relations of the energy industry. These include:

- The 'resource curse' literature which exposes the poor developmental records of resource-rich developing states and the factors which contribute to this, such as the 'Dutch disease' and the failure to develop other sectors of the economy.

- The 'rentier state' and the consolidation of neopatrimonial authoritarian regimes in resource-rich states which are seen to undermine civil society, accentuate the repressive functions of the state, and prohibit the development of democratic states with constitutional restraints on executive power.
- The 'resource wars' which are generated by the predation of natural resources and the breakdown of neopatrimonial states into warring factions whose primary incentive is the capture of rents (the 'greed' rather than 'grievance' explanation for the 'new wars'). This also feeds more generally into the political economy of 'new wars' literature.

Academic and populist syntheses of these various strands can be seen, for example, in the very widely-read *The Bottom Billion* by the former World Bank economist, Paul Collier, who argues that many of the poorest on the planet suffer from living paradoxically in formally resource-rich countries but which suffer from poor economic performance and poor governance (resource predation) and where there is a strong likelihood of their citizens being sucked into debilitating and poverty-inducing civil wars (resource wars). Another populist variant of this is found in the New York columnist, Thomas Friedman who identified the 'First Law of Petropolitics' which is that 'the higher the price of crude oil, the more free speech, free press, fair elections, and independent judiciary, the rule of law and independent political parties are eroded'. This can be seen as the classic statement of the causal linkages between oil wealth and illiberalism.

The more explicitly IR dimension of this liberal exposure of the 'dark underbelly' is in identifying the ways in which Western or other oil-dependent countries, companies and individuals are complicit or alternatively 'turn a blind eye' to these illiberal practices. This is seen in all of its complex manifestations in, for example, an in-depth World Bank study of corruption in the petroleum sector by McPherson and MacSearraigh. Ken Silverstein has exposed some of the shady world of the secret 'oil' fixers, traders and dealers, while another investigative journalist, Nicholas Shaxson, has written about the 'dirty politics' of oil in Africa, describing for example the complex and longstanding 'special relations' between the oil company, ELF, the French Secret Services, and influential parts of the French establishment with the oil-rich state of Gabon and its corrupt leader, Omar Bongo. Robert Vitalis has provided an in-depth historical study of US-Saudi relations and how racist labour relations were transplanted from the US to the ARAMCO-controlled oil fields in Saudi Arabia. These examples of 'secret alliances' and 'special relationships', to use the denunciatory language of liberal internationalism, can be expanded to include many other cases: pre-revolutionary Iran and the US/UK; pre-Chavez Venezuela and the US; Nigeria-UK; Burma and France (and now Burma and China); Russia and Central Asia and the Caucasus; Mobutu's Zaire and France; South Africa prior to the 1990s with the UK and the US...etc

The '*What needs to be done*' provides the reverse side of the '*dark underbelly*': identifying the liberal prescriptions required to overcome the cumulative effects of these illiberal practices and institutions within the international energy realm. These prescriptive policy-driven recommendations come from multiple sources: dispassionate neo-classical economists, activist-scholars, international financial institutions, and campaigning NGOs such as 'Global Witness', 'Revenue Watch' and 'Publish what you Pay'. A good example of this latter category is the International Crisis Group which has an energy security programme which produces multiple reports highlighting how 'from Latin America to the Caucasus and from Africa to the Middle East, energy issues are among the root causes of both civil and interstate conflict'. These reports always include a set of recommendations which the EU, the US, and international institutions are urged to adopt. These normally conform to fairly classical liberal prescriptions:

- *Transparency and the need to develop transparency measures* to avoid the 'secret' deals which underpin the illiberal practices noted above. It is a key liberal principle that openness of information is an essential precondition for ensuring political participation. As Terry Karl notes, 'civil society is useless without information'. The 'Extractive Industries Transparency Initiative' is the most notable international policy instrument in this regard, which was an initiative of the British government, having initially been proposed by the NGO Global Witness and the campaign of the Open Society Institute 'Publish what you Pay'. It has been signed by about 20 countries who have committed themselves to publish oil company payments and government revenues so as to reduce corruption and increase transparency and accountability.
- *International regulation* so as to limit and deter illicit trade and practices which help to foster conflict. A key example of this is the so called 'Kimberley Process' which sought to overcome the problem of

'conflict diamonds' widely seen to be a root cause of conflicts in many parts of Africa, and which required the global diamond industry to commit itself to the introduction of a universal diamond certification process.

- *Corporate Social Responsibility (CSR).* This is a burgeoning area of responsibility for companies which Frynas defines as having three dimensions: a) companies have a responsibility for their impact on society and the natural environment, sometimes beyond that of legal compliance and the liability of individuals; b) that companies have responsibility for the behaviour of others with whom they do business (i.e. supply chains) c) that business needs to manage its relationship with wider society, whether for reasons of commercial viability or to add value to society. International oil and mining companies have been particularly under international pressure in this regard due to their environmental impact and through having frequently to operate in conflict-ridden regions where there is widespread insecurity and conflict, an often disenfranchised sub-national region in conflict with the national government, and local communities which fail to gain the expected benefits from the mining operations. As a consequence, oil and mining companies have been among the leading companies championing CSR. They have developed corporate codes of conduct and social reporting, which include not just IOCs but also NOCs such as Petrobas and India Oil and Kuwait Petroleum. Such companies have also embraced a number of initiatives, such as the UN Global Compact which sets out ten principles in relation to human rights, labour, the environment and anti-corruption. They have also initiated, funded and implemented significant community development schemes.
- *Good governance.* A common theme in these liberal-inspired prescriptive recommendations is that the promotion of good governance is critical if deeply-rooted illiberal practices of 'rent seeking' and the perversions of the 'rentier state' are to be overcome. This involves, at times, a recognition that all the best international efforts at promoting CSR or introducing transparency measures are likely to be impotent unless the national sovereign authority itself has legitimacy and applies principles of good governance. Policy prescriptions for the promotion of good governance can vary from the highly ambitious/idealist: such as Mary Kaldor's demand for 'substantive democracy, [by which] we

mean genuine political equality... a democratic culture of social relations, underpinned by fair, transparent and accountable procedures' to the more narrow specific recommendations about the appropriate ways to manage oil and stabilization funds so as to manage better commodity price fluctuations.

- *Promotion of regional and international energy regimes and institutions.* A common assumption is that the development and enlarged membership of liberal international energy regimes, such as the IEA or the Energy Charter Treaty (rather than the illiberal cartel of OPEC) facilitates international cooperation. It is often suggested that China, for instance, should join the IEA or Russia ratify the ECT. There are often proposals for regional energy cooperative institutions, such as those proposed for South Asia or for North-East Asia to bring together Russia, China and Japan into a mutually beneficial rather than competitive energy relationship (for example, Srivastra and Misra 2007). The EU is itself often a target of this; that its energy policies are insufficiently liberal and institutionalised and that the key solution to its energy security concerns is through further liberalisation of energy markets.
- *Economic liberalisation.* It is fairly obvious but a key liberal assumption underlying these various prescriptions is that most of the illiberal practices and institutions which underlie the multiple conflicts in the international realm derive from imperfect markets. If international energy were liberalised and principles of comparative advantage where properly instituted then energy and minerals would be provided at not only the most economically efficient way but also without the compulsion of geopolitical competition and the conflicts and wars that ensue from that.

Marxist/Radical Approaches

Marxist-inspired radical approaches to IR, such as dependency theory, structuralism and more recently critical IR theory, can themselves be helpfully viewed as seeking to provide substantive critiques of the dominant IR theories of realism and liberalism. For those in this radical tradition, realism is seen to be almost self-evidently flawed since its explanations of international behaviour assume no potential for radical change and thus *explicitly* condone the structural injustices of the international status quo.

Liberalism is more challenging since, like the radical approach, it does offer policies for change and reform, based seemingly on altruistic and benign universal moral principles, but for radicals these policy prescriptions (such as those noted above) actually *implicitly* perpetuate the underlying deeply unjust structures of international power and domination. For those in the classical Marxist tradition, this is because liberalism supports rather than condemns global capitalism. For dependency theorists, economic neo-liberalism only serves to consolidate rather than dismantle the domination of the North and the oppression of the South. For critical IR theorists, liberals take a 'problem-solving' approach, to use Robert Cox's term, which means relying on technical solutions to the resolution of problems (such as Corporate Social Responsibility), rather than asking more profound questions of the moral and political legitimacy of the contemporary international system, such as the radical implications of the privatisation of national welfare in the hands of foreign multinationals.

A good example of a radical approach to international energy issues can be seen in Ray Hinnebusch's recent textbook on *The International Politics of the Middle East.* In this book, Hinnebusch argues that the Middle East economies 'exhibit many of the classic features of dependency' which include dependence on a few basic export commodity, most notably oil; the failure to process these raw materials into finished high-value goods, thus making their economies dependent on the core; the political salience of dependency links between local economically dominant classes and the core, which detaches these elites from the local populations and inhibits the development of national economies; and a western-financed and supported military-security structure which represses challenges to these dependent core-periphery relations. For Hinnebusch, the modern history of the Middle East is a perpetual struggle between local indigenous forms of resistance, such as the radical Arab nationalist anti-imperialist struggles in the 1970s, and Western states and multinationals who have continually succeeded in repressing these attempts at national autonomy, most notably through the triumph of neo-liberalism in the 1980s and the subsequent collapse of Soviet Union which left the US as the unchallenged hegemon in the Middle East.

This radical anti-Western critique of international relations has undoubtedly been a powerful ideology which has inspired much of the drive for independence and autonomy in the Global South. It provided the intellectual foundation for many of the developments of the 1960s and 1970s related to the international oil and mineral industries: the rise of resource nationalism, the policies of nationalisation, the creation of OPEC and other attempts at mineral cartels, the intellectual sources for the Brandt Report and the promotion of NIEO. It has again implicitly underpinned the revival of these policies in the 2000s, if now in the defence of authoritarian state capitalism rather than Marxist or socialist radicalism. International oil and mineral companies find themselves increasingly challenged as conduits of the interests of the core (whether defined as their shareholders or their home states) rather than the national developmental goals of the states in which they operate. In Russia, a narrative has successfully been constructed whereby the privatisation of the oil and gas industries in the 1990s, and the perceived Western support for the liberal oligarch-dominant Russian state of that period, was part of a general Western plot to weaken the power of the Russian state and its traditional great power status role in regional and international politics. Re-nationalisation has thus been popularly been viewed among Russians as a necessary step to the reconstruction and strengthening of the state. Similarly, the Chinese government and the state oil companies see the Western strategies to limit their international energy and mineral investments, such as the enforced CNOOC withdrawal of its bid for UNOCAL, as part of an overarching strategy which seeks to prevent China's 'peaceful rise' and to preserve Western economic and political hegemony. In the South more generally, and particularly with countries with continuing strong memories of imperialism and colonialism, neo-liberal economic prescriptions can very effectively be portrayed as an instrument for continued domination of the North over the South.

Such radical critiques of the way that the liberal international economic order entrenches and helps to perpetuate certain historically constructed economic structures also contributes, in various diverse ways, to some interesting and considerably more nuanced recent contributions on the politics of energy and minerals. These contributions come from diverse disciplinary backgrounds, but with a particular concentration in political and economic geography, and utilise a variety of theories and methodologies – critical theory, post-structuralism and Foulcauldian analysis , action-network theory , global production network analysis. These analyses are only 'radical' in a loose sense and are generally far removed from the rather crude simplifications of the traditional dependency or world systems approaches. But what they do have in common is dissatisfaction with the 'resource curse' or 'resource dependency' approach which, as noted above, is at the centre of the liberal analysis of the sources of conflict in the international energy realm. They generally do not deny that this 'resource dependency' thesis captures some of the causes of resource-related conflict, but they argue that it critically fails to identify some of larger complexities and inter-connections which link energy and minerals to the shifting dynamics of global capitalism.

The main areas of dissatisfaction with the liberal approach exemplified by the 'resource dependency' thesis include:

- That this approach gives too great a causal determinism to oil. In its crude form, you get bald statements by, for example, the influential political scientist, Michael Ross that 'oil hinders democracy' or 'undermines gender equality'. Clearly, Ross's thesis is more complex than that but what this general approach does is provide a simplified notion that it is the very possession of oil which distorts the normal course of development and not a complex set of historical, social and political relations which construct the conditions within which a country seeks to benefit from its resource wealth. It also tends to lead to rather simplistic prescriptions of the quasi-magical powers of 'good governance'. A similar critique can also be given to the rather simplified notions that the struggle over minerals is *the* cause for the millions dead in Central Africa.
- That the emphasis on resource dependency directs too much emphasis on states and particularly on the perceived perversions and maladies of the resource producing states. What a number of these critics note is that there is a striking absence of the firm in these analyses and, if they are included, they are often presented as almost passive actors without the power to resist the resource-producing states. Gavin Bridge provides, in contrast, a highly detailed analysis of the multiple actors who are involved in what he calls the 'hydrocarbon commodity chain' from natural production, to extraction/production, refining, distribution and then finally consumption and carbon capture. Utilising a global production network analysis, he identifies four types of 'oil firms': a) vertically integrated oil companies b)independent producers c) independent transporters, refiners and distributors and d) oil-field service companies. Through identifying these various firms, he seeks to bring back 'agency' to the oil firm, how it is engaged in complex negotiations and bargaining not only with states but also with other

firms. Through this analysis, he highlights how the oil global network extends all the way from the producer to the consumer and involves complex bargaining between the two; and that firms are engaged at all the points on this 'commodity chain' and are themselves in complex negotiations and bargaining with producer states (particularly at the beginning of the chain which is securing the rights to exploration/ production) as well as with other firms. This work naturally draws from an extensive literature on the global oil industry.

- That the liberal approach tends not only to marginalise the role of the firm but also the role of local and sub-national communities. One of the emphases of traditional dependency theory – the salience of 'enclaves' in global capitalism – is something particularly taken up by those studying the oil and gas industry in Africa and elsewhere. In many resource-rich African states, there are small territorial enclaves of mineral extraction, protected by private armies and security firms, in conditions of more generalised civil war and the collapse of the state. The complex social and political realities of these 'enclaves' provide powerful illustrations of the inter-linkages between international and national companies, national states, foreign states, sub-national regions and local communities. The geographer Michael Watts has looked at this with particular attention to the Niger Delta and identifies the following actors, agents and processes: 'as not only the IOCs, NOCs and the service companies but also the petrostates, the engineering companies and the financial groups, the shadow economies (theft, money laundering, drugs, organised crime), the raft of NGOs (human rights, CSR groups, monitoring agencies), the research institutes and lobby groups, the landscape of oil consumption, and not least the oil communities, the military and paramilitary groups, and the social movements which surround the operations of, and the shape and functioning of, the oil industry narrowly construed'. Ricardo Soares de Oliveira, who has also written extensively on oil and politics in West Africa , has done interesting analogous research on the Angolan NOC, Sonogal, and how it has remained an 'enclave' of competence in a state which has otherwise imploded; however, the national company has not unfortunately contributed to national 'capacity building' but has ended up being at the service of the presidency and its rentier ambitions. Dunning and Wirpsa provide a good case study of how the local/

national/international intersect through an analysis of the entry of US oil companies into Columbia in the 1980s; how this triggered local guerrilla attacks on these installations; how this led to a shift in US policy towards providing counter-insurgency support to the Columbian state so as to protect its vital energy supplies from Latin America; and how this only fed into and exacerbated the civil conflict.

- Finally, there is a more ambitious contribution by Timothy Mitchell who seeks to provide an account of the various ways in which fossil fuels have moulded the forms of democracy, or lack of democracy, over the course of the twentieth century. Thus, he contrasts the critical differences between coal, where the labour struggles of miners contributed in Europe and elsewhere to the emergence of the modern welfare state and social democracy, to oil where labour has been more effectively repressed due to the lesser capacity of labour 'to paralyse energy systems and build a more democratic order'. Like other contributors in this broad area, he also criticises the tendency to focus just on the production end of the spectrum and not at the use of fossil fuels and, in this regard, he brings out the peculiarities of the 'carbon-heavy form of middle class American life' and how this helps to define some of the core features of American democracy.

Returning to the Analytical Framework

The three broad theoretical approaches – realism, liberalism and radicalism – are often viewed as contrasting and incommensurate paradigms of understanding international relations. But they can also be seen as potentially complementary since each of these differing theories tends to focus on particular elements and dimensions of the international system and exclude other parts. The theories can therefore potentially be combined in a more syncretic manner and which thereby offers a more holistic, if less parsimonious, conceptualisation of international relations.

This can be illustrated by looking at how these different theories prioritise certain variables of the project's analytical framework at the expense, to a certain extent, of others. Realism tends to prioritise, in terms of independent variables, the salience of the geopolitical distribution of power, the geographical location of resources, and the value of resources (perceived and actual) but tends to give lesser attention to state-company relations and state capacity. Realist theory also tends to emphasise structure rather than agency and is generally sceptical about the transformative power of regional and international institutions and the other intervening variables identified in the analytical framework, with the partial exception of traditional great power diplomacy. In terms of the dependent variables, realism focuses primarily at global and regional geostrategic tensions and conflicts, and the inter-state conflicts that emerge from these, and pays less attention to local and economic/commercial conflicts and to cooperative and collaborative arrangements.

Liberalism, in contrast, focuses considerably more attention and gives greater weight to agency and the transformative potential of the various identified intervening variables – transparency measures, legal frameworks and norms, regulatory and market measures, and the role of regional and international institutions. In terms of the independent variables, priority tends to be given to state capacity and state-company relations, with a lesser explanatory attention given to the geographical location of resources, the geopolitical distribution of power and the value of the resources. In terms of dependent variables, the focus again contrasts with that of realism in that its focus is more at the cooperative/collaborative end of the spectrum, with the roots of conflict seen to reside primarily in the area of domestic/local conditions (the resource curse) rather than at the global and regional level of inter-state conflict. The Marxist/radical tradition has, paradoxically, more congruence with realism than with liberalism in its broad explanatory framework: it again emphasises structure over agency, with particular attention on the geopolitical distribution of power and the geographical location of resources, and is similarly sceptical with realist analyses about the prospects for regional and international cooperation; but, unlike realism, it tends to be more sensitive to state-regime-firm-local linkages and is more ambitious in seeking to identify the connections between the various levels of conflict in the spectrum set out in the dependent variables.

This mapping of the different theoretical traditions to the project's analytical framework is clearly a rather crude exercise which does not do justice to the subtleties and complexities of many of the individual studies within these broad traditions. But what it does at least do is illustrate that the different theories are often seeking to explain differing dimensions of the issue and prioritising different variables for understanding the causes of conflict and cooperation in relation to access to oil, gas and minerals. There is no *a priori* reason that a synthetic inter-theoretical approach cannot be adopted to provide a more holistic conceptualisation of the overarching intellectual framework for the project. This needs also to be assessed in the light of the other theoretical contributions – the economic theories and the comparative politics theories – which are being developed within the project and their potential inter-connections with these IR theoretical traditions.

Connecting to WP3

In terms of looking forward to WP3, one of the key research objectives identified by this work-package is to provide causal explanations of how issues relating to access to oil, gas and minerals migrate from the commercial and economic to the geopolitical and geo-strategic. The IR theories identified above are generally rather weak in identifying such transmission mechanisms and in providing a more dynamic account of how patterns of conflict/cooperation change and are transformed over time. This reflects a common criticism of IR theories that they tend to be ahistorical and too focused on 'current affairs'. This again is rather a crude representation and ignores the main individual accounts which do critically include this.

Two recent theoretical innovations with IR, admittedly drawn from theories first developed in other disciplines, can potentially help in filling this gap. The first is securitization theory, which draws from social constructivism; and the second is historical institutionalism, which draws from historical sociology. Neither of these theories have been explicitly applied to international energy or mineral politics.

Securitization theory was first developed by Barry Buzan and Ole Waever in their book *Security; An New Framework for Analysis* which adopted a social constructivist approach to the study of security by arguing that security threats are not objectively 'out there' in a positivist sense but only come into being through a process of inter-subjective construction. The key intellectual move was to shift away from seeking to determine what security is (the traditional realist security studies approach) to focusing on what security does or how it is constructed. As such, they define security as 'the move that takes politics beyond the established rules of the game and frames the issue either as a special kind of politics or above politics' (p. 23), a process they define as 'securitisation'. In this approach, any public issue can be located on a spectrum ranging from the 'non-politicised' to the 'politicised' to 'securitised'. An issue is defined as being 'securitised' when it is perceived and framed as an existential threat, which underlines its importance and urgency in dealing with it. The central research focus is identifying the process through which something becomes securitised (and de-securitised) and two elements are particularly important here:

- *securitising actors*. Who is it who successfully presents an issue as a security issue? In theory, anyone can be a securitising actor but, in practice, it is those with authority and social power usually derived from their position, who do this successfully and popularise a security discourse. The media can be seen as such an actor as well as political leaders and government bureaucracies.
- *securitization move*. However, not every presentation of something as an existential threat is successful and automatically leads to securitization. An issue is successfully securitised 'only if and when the audience accepts it as such'. They key question is therefore identifying the process by which the broader audience is convinced by certain securitising actors that an issue truly represents an existential threat which puts it above the normal political processes. Examples of this can be seen with various attempts to make AIDS or climate change 'security' issues. Clearly, it also applies in relation to oil, gas and minerals where the perceived threat of 'peak oil' or 'peak minerals' or 'China as a threat to global energy and mineral security' can and are securitised by various securitising actors.

Historical institutionalism as a theoretical approach has a number of areas of convergence with social constructivism in that it is an approach which also seeks to chart the process through which certain critical shifts in international politics emerge and are then incorporated into international practice.

In the IR social constructivist literature, this includes analysis of certain normative transformations in international behaviour, such as the delegitimisation of imperialism, the postwar security culture in Japan, and the emergence of international human rights regimes and the changing norms on intervention. But historical institutionalism also draws from historical sociology, which provides broader macro-historical studies of how certain crucial developments, such as the differing trajectories of states' processes of democratisation, are linked to certain differences in initial conditions, such as state formation processes. In contrast to social constructivism, there is a harder-edge historical materialist dimension to historical sociology.

The particular potential utility of historical institutionalism lies in its ability to offer a more is to offer a more historically sensitive account of the elements of change and continuity over time. These include:

- the role of *critical junctures* and the associated notions of path dependence and positive returns. The notion of critical junctures and the associated model of path dependency is drawn originally from economics; the most famous instance is that of the QWERTY keyboard, which David argued illustrated the ways in which certain technologies can achieve an initial advantage over alternative technologies and can prevail over time despite the potential greater efficiency of these alternatives. The key insight here is the idea that there are certain events, which might be quite seemingly insignificant and contingent at the time, subsequently set in path courses of action which become difficult to reverse.
- The ways in which *positive feedback.* 'lock in' specific developmental trajectories. These involve, as Thelen argues, two processes incorporating self-reinforcing positive feedback. The first which has been particularly developed in Douglass North's work on the history of economic institutions is how, once institutions are created, they establish powerful inducements that reinforce their stability and where actors adapt their strategies in ways which also strengthen and consolidate them. Over time, the dynamics unleashed mean that it becomes more and more difficult to reverse the decisions made at the original critical juncture. The second feedback process is the distributional effects of institutions, which reproduce and magnify particular patterns of power distribution. This involves a recognition, as in neo-realist analyses, that institutions are not just neutral arenas but ones in which certain actors are strengthened relative to others, which in turn consolidates and entrench particular institutional arrangements.
- The analytical focus on institutions. These can be understood in a relatively broad fashion as not just formal institutions, such as OPEC or the IEA, but also as looser regimes and social practices which involve 'sets of rules that stipulate ways in which states should cooperate and compete with each other'. This incorporates potentially much less formal 'institutions' or 'sets of rules' about how states, oil companies and other critical actors engage in the international energy realm.

These two approaches offer some additional potential explanatory application to understanding the transmission mechanisms for how energy- and mineral-related issues can migrate from the commercial and economic to the political and geostrategic realms, a key research objective of WP3.

References

Blank, Stephen. 1995. 'Energy, Economics and Security in Central Asia: Russia and its Rivals.' *Central Asian Survey*, 14 (3), pp. 373–406.

Carr, E. H. 1946. *The Twenty Tear's Crisis: an introduction to the study of International Relations*, London: Macmillan.

Frynas, Jedrzej George. 2009. *Beyond Corporate Social Responsibility. Oil Multinationals and Social Challenges*. Cambridge: Cambridge University Press.

Gilpin, Robert. 1987. *The Political Economy of International Relations*. Princeton: Princeton University Press.

Kaldor, Mary; Terry Lynn Karl and Yahia Said. 2007. *Oil Wars*. London: Pluto Press. Kandiyoti, Rafael. 2007. *Pipelines: Flowing Oil and Crude Politics*. London: I.B. Tauris

Klare, Michael. 2008. *Rising Powers, Shrinking Planet: The New Geopolitics of Energy*. New York: Henry Holt and Company.

Soares de Oliveira, Ricardo. 2007. *Oil and Politics in the Gulf of Guinea*. New York: Columbia University Press.

7

Geopolitics of Energy in the Middle East

The Middle East is a highly diverse area that includes twenty-one nations which are located in an arc that sweeps from North Africa to *the* edge of Central Asia and the Red Sea. These states have a total population of some 280 million, and a GNP of some $578 billion. Most of these states are Arab and Islamic, but they often share little else in political, economic, and strategic terms. Decades of Pan-Arab rhetoric and failed efforts at regional cooperation cannot disguise the fact that the major trading partners of every Middle East state consist of states located outside the region and that the internal character and strategic interest of given nations differ sharply from state to state. As is the case in most parts of the world, the clashes within civilizations are far more important than the clashes between them.

The Middle East is further divided into energy exporters and non-exporters. This creates natural divisions of interest, but energy exporters are also divided into states with large or limited reserves, high or low cost production, and well-developed fields or large potential reserves. All have a common interest in maximizing oil revenues, but for some states this means large-scale exports at low to moderate prices and for others it means trying to restrict the production of other states to increase the value of national exports. Each exporting state has a different regime and different set of interests. Iran, Iraq, and Libya, for example, *are* sometimes seen as radical or threatening states – but they have little in common. Saudi Arabia is a conservative and relatively stable Islamic monarchy that has equally little in common with an Algeria that is in the midst of a bloody civil conflict.

What the Middle Eastern exporting nations do have in common is that virtually all forecasts indicate that their combined output dominates world energy exports and will do so well beyond the year 2020. The Middle East has roughly 64% of the world's proven oil reserves and 34% of its gas reserves. According to estimates by the U.S. Department of Energy (DOE), it exported an average of 17.7 million barrels of oil a day (MMBD) in 1995. This was 47% of the world total of 37.7 MMBD. The DOE projects that Middle Eastern oil exports will reach 39.8 MMBD by 2020. This will be 60% of the estimated world total of 66 MMBD. Similar estimates are not available for gas exports, but Algeria, Libya, Iran, Qatar, Oman, and the UAE will play an important role as world suppliers.

The U.S. is part of a global economy that is dependent on these energy resources. There are many different estimates of the world's future energy balance, but virtually all agree on some broad trends.

- the Department of Energy's reference case indicates that oil will play a dominant role well beyond the year 2020, and that global consumption will grow steadily throughout the period. Rather than running out of oil, the world is running to oil exports.
- both the Department of Energy and International Energy *Agency* estimate that the Middle East will play a critical role in shaping world oil supplies through 2020.
- the estimated size of Middle Eastern oil exports relative to those of other regions in 2020.
- Middle Eastern exports as a percent of the global total, and
- Middle Eastern oil will be distributed to all of the major industrialized states and regions in the world, as well as fuel the development of Asia – the fastest growing region in the developing world.

The issues and uncertainties in such trends will be explored in depth in the analysis that follows, but one central geopolitical reality shapes the demand for Middle Eastern imports. The percentage of oil that flows from the Middle East to the United States or any other country at any given time does not determine the strategic or economic importance of the Middle East or its energy exports to that country. Oil is a global commodity and the U.S. and all other oil importing states must compete for the same global pool of oil exports and pay the same globally-determined price as any other nation. The U.S. and all the other OECD states are required to share all available imports in a crisis under the monitoring of the International Energy Agency. Further, the U.S. economy is particularly dependent on the health of the global economy and on energy-intensive imports from Asia and many other states. In this case, what comes around must go around.

As a result, the most critical single geopolitical issue affecting the region is whether the Middle East will act as a stable supplier of oil and gas exports at market-driven prices. This is not easy to predict in a region that has many intraregional and internal conflicts, serious economic problems, and major demographic problems. The Middle East is so heavily dependent on the income from energy exports that few nations will voluntarily limit their export revenues. War has had a major impact on energy exports in the past, however, and sanctions affect key exporters like Iran, Iraq, and Libya. New questions are also beginning to arise as to whether the Middle East can finance the energy development it needs without more privatization and much higher rates of foreign investment.

Inter and Intra-Regional Issues

The "Middle East" is a regional title that conceals as much as it explains. The nations in the Middle East are divided into at least four sub-regions whose nations often have different interests and present different risks. These four sub-region include the Maghreb, with Mauritania, Morocco, Algeria, Libya, and Tunisia; the Levant and the Arab-Israeli confrontation states: Egypt, Israel, Jordan, Lebanon, and Syria; the Gulf: Iran, Iraq, Kuwait, Bahrain, Qatar, Saudi Arabia, the UAE, and Oman; and the Red Sea states like Yemen, the Sudan, and Somalia. These include states that have been the source of terrorism or conflict, but also many states with a long history of friendship to the West.

It is important to note that Middle Eastern states have strong ties to states outside of the region. The major trading partners of each Middle Eastern state are outside of the region. The nations of North Africa are linked closely to Southern Europe, and also have ties to the Sub-Saharan states. The states of the Levant trade primarily with Europe and the U.S.. The Southern Gulf states trade with the West and increasingly with Asia and the developing world. Iran is in many ways a Central Asian state that exports through the Gulf. It has good reason to be deeply concerned about security issues in Afghanistan and proliferation in India and Pakistan. Treating the Middle East as a "region" rather than as a group of disparate actors often conceals far more than it reveals.

On-Going Violence and the Risk of War

Many Middle Eastern nations have a long history of violence and conflict. The Arab-Israeli Wars of 1948, 1956, 1967, 1970, 1973, 1982, and the Intifada are *all* cases in point. So are the Iran-Iraq War and Gulf War. There *is little chance* that such conflicts will end between the present and 2020. Every Middle Eastern state now disputes at least one border with one of its neighbors, and every country has serious religious and/or ethnic divisions. Low level conflicts and internal unrest are virtual certainties.

In several cases, Middle Eastern states are already at war or confront a serious risk of conflict. Mauritania is the scene of a long *running*, low-level race war between Arabs and Black *Africans*. Morocco is still in the process of an equally long war with the Polisario for control of the Western Sahara. Since 1988, Algeria has been involved in a bitter civil war between its ruling military junta and Islamic extremists. Tensions have grown between Libya's leader, Muammar Qadhafi and Libya's Islamists and there is low-level fighting in a number of areas.

The Egyptian government is at peace with its neighbors but faces a threat from Islamic terrorists. Jordan is *currently* at peace with Israel, but faces a potential threat from Israel and has almost nightly low-level clashes with smugglers and infiltrators from Iraq. The emerging Palestinian entity in the West Bank and Gaza confronts an extremely uncertain peace *while* terrorism and internal clashes continue. In spite of the Arab-Israeli peace process, Israel is still formally at war with Syria and Lebanon, and faces a serious rejectionist threat from terrorists, Iran, and Iraq. Israel is also involved in an active low-level conflict on its northern border with the Hezbollah – a Shi'ite Islamic movement with strong Iranian and Syrian sponsorship. Lebanon remains under Syrian and Israeli occupation, and its factions still present the threat of another round of civil war.

The Southern Gulf states are relatively stable, but tensions exist between Bahrain and Qatar, there is civil violence in Bahrain, and Saudi Arabia and Yemen continue to clash along their common border. While Iran may be becoming more moderate, there is still a serious risk of internal clashes between its "moderates" and "traditionalists," and it presents a major problem in terms of both proliferation and continued hostility to any U.S. presence in the Gulf Iraq remains a serious potential threat to Kuwait and Saudi Arabia, and is certain to resume its military buildup and efforts to proliferate the moment UN sanctions are lifted.

The Red Sea area is the scene of several ongoing conflicts. The Sudanese civil war threatens to enter its second decade, and the dea*th toll* from fighting and starvation probably exceed well over one million. Yemen faces growing tensions between its government and key tribal groups in the South, and continues to clash with Eritrea over the control of islands in the Red Sea.

External tensions are compounded by internal problems. Most Middle East states suffer from internal political, economic, and demographic problems that *exacerbate* these intra-regional conflicts and tensions. Virtually all Middle East states have repressive regimes with a high degree of authoritarianism – regardless of whether the ruler is called a King, Sheik, Sultan, President, General, or Ayatollah. Virtually all suffer from weak or failed economic development *programs*, high rates of population growth and a virtual youth explosion, aging and largely authoritarian regimes, and serious problems with internal stability.

The Middle East also suffers from a process of creeping proliferation that may ultimately change the nature of conflict and the balance of power in the region. Algeria, Egypt, Iraq, Iran Israel, Libya, Syria, and Yemen have all created missile programs and have at least conducted research into weapons of mass destruction. Israel is a major regional nuclear power and has chemical and biological programs. Egypt has chemical and biological programs. Iran, Iraq, Libya, and Syria *are* either *developing* biological and chemical weapons *or already producing them*, and Iran and Iraq continue to seek nuclear weapons. So far, such weapons have only been used in the Yemeni civil war and the Iran-Iraq War, but there is little doubt that the Middle East is acquiring far more lethal systems than it has possessed in the past.

Militarism and State Violence

External tensions and risks are only part of the problem. Many of the regimes in the region are repressive, and state terrorism is endemic. The U.S. State Department reports that the majority of Middle Eastern states were guilty of acts of state terrorism last year. Several states actively sponsor external terrorist movements or conduct acts of terrorism outside their own territory. These states include Iran, Iraq, Libya, and Syria. Other states face serious terrorist threats. These states include Algeria, Bahrain, Egypt, Israel, Jordan, Lebanon, and Saudi Arabia.

Militarism remains a serious problem, although there has been a sharp decline in regional military expenditures and arms imports since the end of the Cold War. Middle Eastern military expenditures dropped from $93.0 billion in 1985 to $48.6 billion in 1995, measured in constant 1995 U.S. dollars. Middle Eastern arms imports dropped from $27.9 billion in 1985 to $48.6 billion. North African military expenditures dropped from $8.3 billion in 1985 to $13.8 billion in 1995. North African arms imports dropped from $3.5 billion in 1985 to $320 million in 1995.

This decline in arms imports and military expenditures has not been a matter of choice. It has been enforced by the end *of* concessionary arms transfers by the FSU, growing regional economic problems, and a range of sanctions on key states like Iran, Iraq, and Libya.

Furthermore, the Middle East is so heavily armed that the drop in arms imports has *had* little substantive impact on war fighting capability. The region still spends nearly 8% of its GNP on military expenditures, which compares with an average of 2.8% for both the developed and developing world.

These pressures have led some nations to shift their resources away from conventional forces to the acquisition of weapons of mass destruction, long-range delivery systems, and carefully selected advanced conventional weapons. At the same time, many countries have had to greatly expand their internal security efforts, creating new risks of regional instability and terrorism.

Economic Background to Geopolitics

Economic policies have also been a threat to regional stability. Regional economic development has lagged since the end of the oil boom in the late 1970s. The Middle East only averaged 0.2% annual economic growth from 1980-1990, only 6% of its average annual population growth. This situation has improved *slightly* since 1990, but growth averaged less than 3% before the economic collapse in Asia and *the* similar collapse in world oil prices in late 1997. Population growth slightly outpaced real economic growth throughout the 1990s.

Some states like Kuwait, Qatar, and the UAE have so much oil and gas wealth per capita that they can still buy their way out of their mistakes. Most Middle Eastern oil exporting states, however, face a growing inability to fund their domestic programs, and suffer severely from economic mismanagement and excess*ive* state control of *their economies.* Structural economic reform has begun in Algeria, Morocco, Tunisia, Egypt, Jordan, Lebanon, and Bahrain. The progress of such reforms, however, remains highly uncertain and Egypt is the only state *undergoing such reform* that has carried it forward to the point where it has a serious prospect of success.

Many Middle Eastern states have very uncertain near to mid-term economic prospects, and this is true of most oil exporters. The Iraqi economy is in near collapse. The Iranian economy is in a serious crisis, compounded by deep ideological conflicts over how to deal with the issue. Algeria's efforts at reform are blocked by civil war. Qadhafi's mismanagement and UN sanctions have blocked much of Libya's development. Bahrain no longer has significant oil reserves. Saudi Arabia has experienced over a decade of budget deficits and now has less than 40% of *the* real per capita income it had at the peak of the oil boom. Oman is *also* experiencing serious development problems

Demographics and Geopolitics

Ineffective population programs and a "youth explosion" compound the impact of failed leadership and economic mismanagement. Population growth still averaged 2.7% during 1995-1996, and exceeded 3.4% during the period from the late 1960s to 1990. The region experienced negative real economic growth during much of the 1980s, and economic growth only averaged about one-third of population growth during the 1990s – before the collapse of oil prices in 1997. The *region's* average per capita income dropped by 1.8% during 1995-1996.

Roughly 40% of the region's population is now under 17 years of age. The region's educational system is under extreme stress, and real and disguised unemployment for males between 18 and 25 years probably averages over 20%. The average per capita income of the Middle East is now about $2,070 using the World Bank method, and compares with $25,890 for high-income states. Urbanization without development, proper infrastructure, and housing has *intensified* the region's problems. The percent of urbanization in the total population rose from 41% in 1971 to 58% in 1997.

This combination of low oil prices, high population growth rates, and a failure to modernize and diversify the overall economy threatens to turn oil wealth into oil poverty. The Southern Gulf states have only about 40%

of the real per capita income *that* they had at the peak of the oil boom in the early 1980s, and *little hope* for anything other than a slow decline. Kuwait, Qatar, and the UAE *maintain* high per capita incomes, but Saudi Arabia is becoming increasingly marginal, Iran has a per capita income of $1,780, Algeria has $1,500 dollars, and Iraq's per capita income is unlikely to be higher.

Pressures for Geopolitical Change

The Middle East has to undergo fundamental changes by 2020. A stable future depends on key political events like the success of the Arab-Israeli peace process, the moderation of the Iranian revolution, the creation of a stable and peaceful Iraq, and an end to the civil war in Algeria. Most of *the* key leaders are aging and the highly personal patriarchal systems of government they have established are unlikely to survive them. *Regional Stability* depends on the succession of stable and more progress*ive* regimes in key states like Bahrain, Egypt, Jordan, Libya, Kuwait, Morocco, Oman, Saudi Arabia, and the UAE.

A stable future also depends, however, on the ability of the region to implement the level of energy development that it, and the world, need over the coming decades. There are strong indications that this will require major changes in the way most countries finance their energy development and in the role of foreign investment. These issues are discussed in the energy sections of this analysis.

Finally, even the most successful development of the region's energy resources is unlikely to fully meet its economic needs and bring regional stability. A stable future requires the willingness of exporting states to implement sustained economic reform, to come to grips with the need to reduce population growth, and *to* reduce dependence on foreign labor. The region must either further reduce its population growth rate or breed itself into poverty. Similarly, every oil-exporting state must improve the management of their economies, diversify, and shift to far greater reliance on free markets, or risk economic collapse.

These pressures can create many possible futures. One alternative is that the region will do enough to "muddle through," largely preserving the status quo. In this case, political change will be limited and economic growth will barely keep up with population growth, if at all. Internal tensions will grow worse, civil conflict will continue, and some low-level fighting will take place between states.

Another *possible* future is that most exporting states will react in a positive manner, and will steadily implement the reforms they need in ways that avoid sudden political and domestic shocks. This kind of future is still open to the Southern Gulf oil exporting states and Libya. The political, demographic, and economic pressures on Algeria, Iran, Iraq, and Syria are so great that even partial success – "muddling through" – may be difficult.

The high-risk scenario involves a mix of different *variables* that can interact in very unpredictable ways. These *variables* include a succession crisis or internal instability in Saudi Arabia, instability in Iran and a transfer of power to revolutionary extremists, and continued revanchism and authoritarianism in Iraq. They *also* include the problems in the Arab-Israel peace process, and continuing civil conflict in Algeria and Libya.

The most likely future is a combination of all these futures that varies sharply by country. Most Middle Eastern states have, after all, muddled through for decades with considerable success. Many "problem states" have remained problems for the same period. A number of Middle Eastern states are seriously debating reforms and beginning to implement them. The fact that there are many risks and problems is by no means a prophecy that most things that can go wrong will go wrong.

The problems and risks do mean, however, that there are many geopolitical issues in the Middle East that affect Western interests and the global economy. Energy geopolitics, however, are dictated in large part by which state and subregion produces the most oil and has the largest oil reserves. From this viewpoint, the most important of these *issues* involves the stability of the Gulf and *the* key energy exporters in the Gulf, the stability of Algeria and Libya, and the extent to which external factors like the Arab-Israeli conflict affect energy developments.

The Gulf has long dominated regional oil production, and that North Africa ranks second. The oil production from the states in the Levant is important in meeting the domestic demands of Egypt and Syria, but is too small to have any significant impact on world supplies.The Gulf dominates not only Middle Eastern b*ut also* world proven oil reserves, followed by North Africa.

- The pronounced dip in the mid-1980s reflects the impact of the "oil crash" of 1986. The irregularities in Iranian, Kuwaiti and Iraqi production, and surges in Saudi production, reflect the impact of the Iranian revolution, Iran-Iraq War, and Gulf War. Geopolitics and economics clearly interact to shape regional production.
- It to some extent reflects the drop in production caused by the "oil crash" of 1997, but the behavior pattern is different. In spite of efforts to reduce production to raise prices, Middle Eastern states are now so desperate for cash flow that they cannot make sharp cuts in production, *so they* tend to cheat on quota agreements. Another important trend is the reemergence of Iraq as a major producer following its "oil for food " deal with the UN.
- In spite of more than a quarter of a century of effort to find oil reserves outside the Middle East following the oil embargo of 1974, the Middle East now *has* a larger percentage of the world's proven oil reserves than it did in the 1970s. It is also important to note that the leveling out of new discoveries is heavily impacted by Saudi Arabia's decision to limit data on new reserves, and by war, sanctions, and internal turmoil in Algeria, Iran, Iraq, Kuwait, and Libya. As will be discussed later in this analysis, the potential *discovery* rate in the Middle East remains extremely high.
- the Gulf dominates world holdings of proven oil reserves, with approximately 65%. The rest of the Middle East holds another 5%, with 4% in North Africa (Algeria and Libya), and 1% in the Levant (largely Egypt and Syria.)
- the dominant role of Saudi Arabia, followed by Iraq, the UAE, Iran and Kuwait. Note that Levantine and North African states have a much lower share of proven reserves than of current production, a fact that reflects their declining mid to long-term geopolitical importance as oil producers.

Once again, it should be stressed that the broad trends shown in these figures illustrate geopolitical realities over which there is little debate. The specific estimates, however, involve important uncertainties that will be discussed in the sections on energy. They also ignore gas, which is another critical aspect of domestic energy production and regional reserves. At present, Middle East gas exports are relatively limited, and are dominated by Algeria, with Oman and Qatar emerging as exporters to Asia. Gas exports are, however, becoming an important issue *in* Iran, and there is a possibility that advances in technology will lead to changes in the cost of gas liquids, *which* could make gas more of a geopolitical factor in the future.

Trends in the Gulf Countries

While many studies talk about the oil wealth of the "Middle East," most of the region's energy reserves and production are concentrated in the Gulf. The Middle East as a whole may have more than 65% of the world's proven oil reserves and 40% of its gas reserves, but over 90% of these oil reserves are in the Gulf. *Similarly, the U.S. Department of Energy (DOE) estimates that the Gulf averaged 15.4 MMBD worth of exports in 1995, which equaled 41% of all world exports versus 47% for the entire Middle East. DOE estimates that the Gulf will average 37.2 MMBD in exports by 2020, which will equal 56% of all world exports compared to 60% for the entire Middle East.*

Once again, there are important uncertainties in these estimates. *Given the* DOE's reference case *for* estimates of Gulf production and the probable range of uncertainty, Gulf production will increase from 18.7 million barrels per day in 1990, and 22.8 million barrels per day in 1997, to 23.9 (23.5-24.2) million barrels per day in 2000, 28.1 (25.8-30.2) million barrels per day in 2005, 29.6 (27.1-35.4) million barrels per day in 2010, 34.9 (28.6-42.8) million barrels per day in 2015, and 42.2 (35.3-52.3) million barrels per day in 2020. This is a potential increase from 28% of all world production in 1996 to 42% in 2020.

Many experts in the oil industry dispute these numbers. Few, however, dispute the central geopolitical risk involved. The Gulf will become steadily more important with time, not only as a percentage of total production and exports, but also as a "swing" producer whose ability to increase production and exports will stabilize world prices.

The Gulf also has major gas reserves, and is becoming a major exporter of liquid natural gas. While the Russian Federation dominates the world's reserves with 1,700 trillion cubic feet, or 33.4% of the world total, Iran alone has 16% of the word's gas reserves and Qatar and the UAE have another 10%. In total, the Gulf has over 33% of the world's reserves. The rest of the Middle East adds less than another 3%.

Saudi Arabia

Saudi Arabia has become one of the largest and most powerful states in the Gulf, and a key supplier of world oil imports. In 1998, Saudi Arabia had well over 260 billion barrels of proven oil reserves, or one-quarter of the world's total. Saudi Arabia is the pivotal oil exporter in the Gulf, the Middle East, and the world. It is the world's largest "swing" producer, and plays a critical role in ensuring moderate and stable oil prices.

Saudi Arabia is the world's largest oil producer, and the growth in Saudi oil production will outstrip the growth in all of the nations in the Former Soviet Union, in spite of major increases in production by the former Soviet republics in the Caspian and Central Asia. *The U.S. Department of Energy estimates that Saudi Arabia will increase its production capacity from 11.4 million barrels per day in 1997, to 20.0 million barrels per day in 2020. The Department of Energy estimates that Saudi Arabia's production will shift from 14.7% of world production in 1997, to 13.8% in 2000, 14.8% in 2010, and then rise to 17.8% in 2020.*

Saudi Arabia has long had close security ties to the U.S., and has been a reliable exporter of oil since the end of the oil embargo of 1974. It is the only Southern Gulf power strong enough to serve as a strategic counterbalance to Iran and Iraq, and is a critical partner in U.S. collective security efforts in the region.

Political, Economic, and Social Transition

Saudi Arabia does not face any imminent risk of instability, but it will enter the twenty-first century in the midst of major political, social, economic, and military transitions. External transitions include the reemergence of Iraq as a major force in Gulf security and the world oil market, Iran's uncertain shift towards political moderation and regional cooperation, the failure of the Southern Gulf states to develop their military forces effectively and develop meaningful collective security arrangements, creeping proliferation, and the need to redefine dependence on the U.S. for security. They also include the continuing uncertainties in the world energy market, a factor that drives virtually every aspect of the Saudi economy.

The most visible internal transition will be political: King Fahd's health continues to deteriorate and the Saudi royal family must agree on a successor. For the first time in decades, Saudi Arabia is likely to have a king who is not one of the "Sudairi Seven." Almost inevitably, this "succession issue" has focused the world's attention on whether Prince Abdullah will come to power, on how the Saudi regime will change, and on whether there will be major changes in its foreign and domestic policy.

Yet, it is the social, economic, and military transitions in Saudi Arabia that may have a much more lasting effect. In 1973, before the beginning of the oil boom, Saudi Arabia was a nation of roughly 6.8 million people. It had a GNP of less than $10 billion in market prices, and a per capita income of less than $2,500. It was largely rural or nomadic and largely pre-industrial in character. While no precise figures are available, its population growth rate was probably under 2.7% and less than 30% of its population was under 14 years of age.

The oil boom that began in 1974 transformed Saudi Arabia. By 1997, it was a nation of 20.1 million people. Its economy had become so different that it is impossible to make direct comparisons of its GNP before and after the oil boom. T*he World Bank estimated that in 1997, Saudi Arabia's GNP was $143.4 billion, and that Saudi Arabia's per capita income was over $7,000.* The CIA estimated that its GNP was $189.3 billion and that its per capita income was $10,100. The CIA estimates for 1996 were a GDP of $198 billion with a per capita GDP of $10,200, and the estimates for 1997 were a GDP of $206.5 billion with a per capita GDP of $10,300. In contrast, the EIU estimated a GDP of $135.5 billion for 1996, and $139.6 billion for 1997.

It is clear, however, that oil wealth has transformed a poor and largely agricultural Saudi Arabia into a heavily urbanized, welfare state with a large service sector. Agriculture shrunk to only 9% of the GDP by 1995, while industry rose to 50% and services to 41%. The labor force shifted from a time when 64% of the total worked in agriculture to a force where only 5% works in agriculture while 40% works in government, 25% works in industry and oil, and 30% works in services. Equally important, a subsistence society has become a welfare society where *close to* 20% of all personal income comes from government grants, services, and subsidies.

Much of this development *has had* positive *effects*. Oil wealth allowed Saudi Arabia to spend some $1,124 billion dollars on development between 1970 and 1995. Saudi Arabia had 8,000 kilometers of paved roads in 1970, and 67,893 in 1995. It had 27 wharves at its seaports in 1970 and 182 in 1995. It had 3,283 elementary schools in 1970 and 21,854 in 1995. It tripled the number of hospitals, set up 3,300 community health centers. It increased the number of doctors at its hospitals and clinics from 1,172 to 30,306, and its nursing staff from 3,261 nurses to 60,736 nurses and 33,047 technicians. It increased the output from its desalination plants from 5.1 million gallons to 512 million gallons, and its electric generation capacity from 344 megawatts to more than 17,400 megawatts. Saudi Arabia had 119 factories in 1970 and 2,303 in 1995, with an invested capital of $40.7 billion.

Internal versus External Threats to Saudi Security

At the same time, Saudi Arabia now faces many challenges. Its military forces are not strong enough to defend the Kingdom from Iran and Iraq without U.S. military aid, and it is dependent on a strategic partnership with the U.S. and other Western states. It has not been able to catalyze effective collective security efforts within the Gulf Cooperation Council (GCC) and it still faces serious problems in eliminating its historical rivalries with other Southern Gulf states.

Most importantly, Saudi Arabia must deal with the social and economic impacts of explosive population growth at a time when its oil wealth has diminished sharply in both absolute and relative terms. These problems did not begin with the "oil crash" in 1997. The same wealth that transformed other aspects of Saudi society helped raise the average population growth rate to 4.7% during 1980-1995. Although this growth rate declined to an average of 3.3% during 1990-1995, it had risen back to 3.45% in 1995 and grew again in 1996 to 3.6%. While it dropped to 3.42% in 1997, this was still higher than in 1990-1995, and one of the highest rates of growth in the world. The end result is that 43% of the population was under 14 years of age by 1997. The World Bank forecasts that Saudi Arabia's population will grow by about 3.3% per year over the next five years. As a result, even conservative World Bank estimates project a total Saudi population approaching 31 million people in 2010.

Oil wealth also led to radical changes in Saudi Arabia's social structure. Urbanization reached 67% of the total population by 1980 and 79% by 1995 — a total of roughly 14.9 million people. By 1995, over 20% of Saudi Arabia's population lived in cities of over one million — a total of roughly 14.9 million people. Education changed radically: Only 61% of school age Saudi males ever entered primary school when the oil boom began while the percentage rose to over 80% by 1990. The percentage for females rose from 29% to 73%. Literacy in the population as a whole rose from well under 15% of the population to over 60%, and Saudi society became exposed to the world's media. Saudi Arabia had over 250 television sets per 1,000 people in 1997, and over 95% of the Saudi people had exposure to radio.

Structural Economic Problems and the 'Oil Crash'

Since the early 1980s, however, Saudi Arabia's oil wealth has declined in both absolute and per capita terms. Saudi Arabia's population has grown explosively while oil *revenues* have remained nearly constant. The World Bank estimates that Saudi Arabia's population rose from roughly 9 million in 1980, to 19 million in 1995 — a rise of 111%. In contrast, the World Bank estimates that Saudi Arabia's GDP dropped from $156.5 billion in 1980, in current dollars, to $125.5 billion in 1995. This is a drop of nearly 20% in current dollars and well over 30% in constant dollars. U.S. estimates are similar. They indicate that the Saudi GNP dropped by over 35% during the same period.

Furthermore, Saudi Arabia failed to adequately diversify its economy. The *earnings of* its petroleum product and downstream operations *have* been *partly* offset by their cost and the diversion of crude oil to feedstock. Domestic energy use has been extremely wasteful, and much of Saudi Arabia's GDP now consists of service industries whose only real function is to increase and meet the demand for imports. Trade makes up nearly 50% of the Saudi GDP today, but only a few percent of Saudi exports are manufactured. Trade is virtually all in petroleum-related exports and in imports financed by these exports. While estimates differ, virtually all outside analyses agree that Saudi Arabia's per capita income has declined to less than 40% of its peak at the height of the oil boom.

Saudi Demographics and the 'Youth Explosion'

The World Bank estimates that Saudi Arabia's per capita income dropped by an annual average of 2.9% during 1970-1995, and by a total of nearly 20% during 1985-1995. During the same period, however, total Saudi Arabian private consumption rose from $34.5 billion to $52.0 billion. This growth in consumption reflected both the impact of population growth and a growing social dependence on imports and services. Private consumption rose from 22% of the GDP in 1980 to 43% in 1995, while government consumption rose from 16% to 27%. At the same time, gross domestic investment dropped slightly from 22% to 20%, and gross domestic savings dropped precipitously from 62% of GDP to 30%.

Most of Saudi Arabia's young population is not educated for real-world jobs or to compete in the modern world economy and nearly half of its labor force is foreign. While over 90% of Saudi young males and females reached grade four in 1995, only 5% of males and 2% of females moved on to secondary school. Much of Saudi education is "Islamic," and oriented to rote learning, rather than focused on training students for real world jobs and competitive with the leading economies of the developing world. Saudi Arabia educates women with no clear idea of the role they should play in the labor force, although female workers rose from 5% of the total labor force in 1970 to over 13% in 1997. Dependence on welfare, meaningless government and government-related jobs, and foreign labor has left much of native Saudi youth without a work ethic, while declining per capita oil revenues mean the Kingdom cannot hope to sustain its past pattern of disguised unemployment and subsidies.

Transformation is the Price of Stability

Recent increases in education spending will do little to change the character and quality of Saudi education, far too much of which is unfocused, religious in character, reliant on route learning, and provided by low-wage contract teachers from outside Saudi Arabia. Saudi youth have very uncertain prospects of finding real, productive employment unless Saudi Arabia radically restructures its educational system and economy. Its government has swollen to the point where it consumes 40% of Saudi Arabia's labor force, and more than half of its native labor force. The Saudi government costs over $50 billion a year, government spending accounts for 36% of GDP, and is in its fifteenth year of continuous budget deficits.

During the coming decades, Saudi Arabia must make massive new investments to maintain its status as a petroleum power, *and make* equally massive investments in the national infrastructure necessary to support its rapidly growing population. These investments almost certainly will cost in excess of one-third of a trillion dollars, is a far larger sum than Saudi Arabia can self-finance, even if it were to shift from almost complete dependence on state investment to dependence on private financing. Saudi Arabia must restructure its economy to deal with declining per capita oil wealth and the need to rely on native, rather than foreign labor. It must transform its society to compete in a world where oil wealth alone is not enough to keep its per capita income from dropping steadily lower, and do so without losing its Islamic and Arab character.

Saudi Arabia, Iran, Iraq, and Yemen

Saudi Arabia may be able to meet these challenges, but the task will not be easy. Saudi Arabia faces external threats from Iran and Iraq. Saudi relations improved with Iran after the election of President Khatami in 1997, but tensions still exist along sectarian lines and the two states are natural political rivals for power and influence in the Gulf. Iran's efforts to proliferate *present* a major new potential threat to Saudi Arabia, and there are natural differences in *the* energy strateg*ies of the two countries*. Saudi Arabia is a low cost producer with immense reserves that can afford to produce at very high levels even when oil prices are low. Iran has much higher production and development costs, more limited production capacity and reserves, and much less capability to pursue a "market share " strategy. Put differently, Iran benefits from limits on Saudi production.

Iraq is a hostile dictatorship that has repeated*ly* threaten*ed* Saudi Arabia and Kuwait since the Gulf War. It too has major oil reserves, but nearly two decades of war and a decade of sanctions have severely limited Iraq's ability to develop its oil wealth and it needs major investment to rework and modernize its existing fields. It may be years before it can exploit its new fields and costs are difficult to estimate. In the short to medium term, Iraq benefits from limits on Saudi production and needs to take every possible step to maximize its oil revenues. Like Iran, it is a major proliferator and is likely to create major *military* forces the moment UN sanctions are lifted or can be ignored. It is also a major conventional power, and unlike Iran, it shares a common and poorly defended border with Saudi Arabia.

A Saudi border dispute with Yemen dating back 60 years is another possible source of conflict. Yemen has long disputed Saudi Arabia's claim to three Red Sea islands and parts of the Rub al-Khali (Empty Quarter), a desert region thought to contain valuable oil fields. Over the years, fighting has erupted along their 1,300-mile frontier. Tensions rose once again in July 1998 when an armed clash broke out on the Red Sea Island of Duwaima off the coast of Yemen. The fighting lasted for five hours until a cease-fire was negotiated. Each side accused the other of initiating this conflict. *On July 28, 1998*, the two countries agreed to evacuate disputed areas until a solution could be found to the border conflict. This, however, is simply one more agreement in over thirty years of negotiating efforts that have sought to create a stable relationship and border settlement and which have all decayed into renewed tension and conflict.

Saudi Arabia, Collective Security in the Southern Gulf

Saudi Arabia has good relations with Bahrain, and correct relations with the other Southern Gulf states. Although Saudi Arabia has made a serious effort to improve its relations with Kuwait, Oman, Qatar, and Yemen, change comes slowly in the Gulf. Restructuring Saudi society and the Saudi economy presents the most serious challenge of all, *since* change must come at a time when Islamic extremism is a serious concern and Saudi Arabia's oil income is not sufficient to meet its social and economic needs.

Collective security efforts in the Southern Gulf do not protect Saudi Arabia or any other Gulf state. The Gulf Cooperation Council provides a useful political and security forum, and some progress has been made in holding common military exercises and in dealing with internal security issues.

At this point in time, however, both Saudi security and that of every other Gulf state is dependent on bilateral security agreements with the U.S., the U.S. forces deployed in the Gulf, and U.S. power projection capability. At the same time Saudi Arabia's strategic partnership with the U.S. and other Western states involves inevitable differences in national interest and culture, and its partnership with the U.S. presents problems in burden-sharing, counter-terrorism, prepositioning, and longer term problems in defining a stable and equitable division of effort.

Role of U.S. Power Projection and the 'Four Cornered' Balance of Power in the Gulf

The practical balance of power in the Gulf is essentially a four corned structure in which Iran and Iraq compete for regional influence and/or dominance against each other. This competition is counterbalanced by Saudi Arabia, which is too weak to act alone but which can rely on the U.S. for the extra margin of force it needs to provide both deterrence and defense.

This is an awkard juggling contest for all four major players and it is further complicated by the impact of the other Southern Gulf states and a long list of peripheral players ranging from Afghanistan to Russia.

The game also changes with time. Saudi Arabia cannot afford to be too dependent upon the U.S., and the smaller Southern Gulf states have their own fears of Saudi dominance. The lower Gulf states tend to tilt towards Iraq because of their fears of Iran, *while* the upper Gulf states *tend* to tilt against Iraq in favor of Iran. No player, however, can firmly trust any of the others and alliances tend to shift against the greatest perceived threat. Similarly, the U.S. and other Western powers like Britain and France are most popular in the Southern Gulf where they are most needed, but they can scarcely count on "gratitude" when this threatens internal stability and involves a major burden sharing effort. There is a natural tendency to push the U.S. "over-the-horizon" when tensions weaken.

The future security of Saudi Arabia will be determined primarily by its internal security and development. At the same time, it will be determined by the Kingdom's ability to limit Iranian and Iraqi hostility and to balance one Northern Gulf state against the other in ways that enhance Saudi security. This regional balance of power will also require the Kingdom to carefully balance its security ties to the U.S. and the West in ways which provide sufficient security guarantees in terms of forward presence and power projection capability while minimizing the destabilizing effects of a U.S. military presence in the Kingdom, the resulting tensions with its neighbors, the backlash from ties to Israel's closest ally, and the impact of a faltering and uncertain Arab-Israeli peace process.

Iran

Iran is a nation that is still deeply in the process of revolutionary change, and which is deeply divided between "moderates " who have broad public support, and "conservatives" who control the military, security system, and most other governmental institutions. The "moderates" now seem to be the strongest faction, and change may take a peaceful and positive course. Iran's regime has become steadily more pragmatic under President Rafsanjani and President Khatami, and more concerned with Iran's national interests and economic development in the Gulf than the export of revolution. Since President Khatami's election, there have been growing signs that Iran may evolve a more tolerant approach to defining an Islamic state, one that emphasizes the humanitarian and moral strength of Islam, rather than the *attempt* to force other nations into accepting its concept of repressive and outdated theological rule and social customs.

Revolutions, however, can become more extreme as well as more moderate. Iran's pragmatists and moderates still face strong traditionalist and radical opposition. Iran's revolution may yet become the captive of ambitious leaders or elites. Conservative or extremist reaction can suppress the positive trends in political and social development, and nationalism and regional ambition can turn ideology into an excuse for aggression. Economic failure can also become an excuse for aggression, as can the need to justify authoritarian rule and social repression.

Iran's Evolving Foreign Policy

Iran's foreign policy is in transition, hopefully to a moderate state that concentrates on development and *considers* security largely in defensive terms. Iran still has its hard-liners, however, and there have been no reductions in its efforts to proliferate and build-up its military capabilities in the Southern Gulf. Iran has reduced its support for terrorism in the Southern Gulf, and its involvement in incidents has declined since Khatami's election. *However*, Iran's training facilities and infrastructure remain intact.

Iran has made progress in improving relations with a variety of countries. In December 1997, Iran hosted the Organization of the Islamic Conference (OIC) in Tehran. During this well-attended (more than 50 countries) meeting, Iranian President Khatami met twice with Saudi Crown Prince Abdullah, the first such high-level meetings between Iranian and Saudi leaders since the 1979 Iranian Revolution. The meetings led to steadily better relations between the two countries. In February 1998, former President Rafsanjani visited Saudi Arabia for 10 days for talks on *improving* bilateral ties and formulating a "security and economic strategy" for boosting security in the region. Rafsanjani was the most senior Iranian to visit Saudi Arabia since the 1979 Iranian Revolution. Iraninan and Saudi relations continued to improve in 1998 and 1999, including a visit by President Khatami to Saudi Arabia in May 1999.

Iran has also improved relations with Kuwait, Bahrain, and Oman. The key remaining problem area in Iranian relations with the Southern Gulf states is the UAE. Iran seized Abu Musa and the Greater and Lesser Tunbs from *the emirate of* Ras al-Khaimah in 1971. Iran claimed full sovereignty, but reached a face-saving agreement with the UAE to "share" Abu Musa. In 1992, Iran claimed sovereignty over Abu Musa despite the 1971 agreement between the two countries. *The UAE* still maintained joint control of Abu Musa until 1994, when Iran took full control of the island. Negotiations with the United Arab Emirates (UAE) over Abu Musa and *the* Tunb Islands have remained stalled ever since.

The Iranian Foreign Ministry issued a statement in December 1995 declaring that the islands are "an inseparable part of Iran." In 1996, Iran took further *steps* to strengthen its hold on the disputed islands. These moves included starting-up a power plant on Greater Tunb, opening an airport on Abu Musa, and planning the construction of a new port on Abu Musa. Iran also rejected a proposal by the Gulf Cooperation Council in March 1996, which advocated that the International Court of Justice resolve the dispute, an option supported by the UAE.

The UAE has generally received strong support from the GCC, *the* United Nations and the United States, although the UAE and Saudi Arabia have quarreled over Saudi Arabia's improving relations with Iran in May-June 1999. In December 1997, the UAE called for talks with Iran over the islands. In early March 1998, the GCC, while praising Iran's President Khatami, issued a statement supporting the UAE in its dispute with Iran over Abu Musa and the Tunbs. Since that time, however, Iran has shown that it is at least willing to discuss the issue with the UAE, and the foreign ministers of the two countries have exchanged visits. It is also clear that the UAE is not *completely* unified on the issue. *The Emirate of* Dubai is one of Iran's most important trading partners, and Abu Dhabi and Ras al-Khaimah that are responsible for much of the rhetoric.

Iran and Iraq opened a dialogue in 1997 that led to an exchange of the remains of 75 soldiers killed on both sides during the Iran-Iraq War in April 1997. Since that time Iran and Iraq have agreed to exchange all prisoners, and establish more normal relations, and Iraq allowed Iranian pilgrims to visit Shi'ite religious shrines in Iraq for the first time since 1979. Iran has also sporadically assisted Iraq in smuggling refined products, mainly diesel fuel, to international markets. Relations, however, are still anything but friendly. Major tensions exist over Iraq's treatment of its Shi'ites and the assassinations of three major ayatollahs in Iraq during 1998-1999. Both nations deploy a substantial portion of their military forces to defend against an attack by the other state, both support armed opposition movements to the other's regime on their territory, and both *have lasting memories of* the Iran-Iraq War and are involved in a race to proliferate.

It may be a decade or more before Iran's ultimate political course is clear, and it is difficult for many observers to face the fact that it will take time and patience to observe the outcome. In the interim, Iran has developed many critics and apologists. There are those who "demonize" every Iranian action and event, even when such action appears positive or largely defensive. *Similarly,* there are those who "sanctify" Iran's worst mistakes, just as there have been those who have excused or glorified every authoritarian and repressive regime of the Twentieth Century.

Iran and Military Threats in the Gulf

Much of Iran's military behavior is explained by the continuing threat from Iraq, its fear of U.S. intervention, and its desire to play a major – if not dominant – military role in the region. Iraq's military capabilities may *have* been greatly weakened by the Gulf War, but Iraq remains a stronger military power than Iran. Iraq has roughly 380,000-400,000 men under arms versus 345,000 men in both Iran's regular forces and the Revolutionary Guards. Iraq has 2,700 main battle tanks and about 3,800 other armored vehicles compared to less than 1,400 and 1,100 respectively for Iran. The only category of major land weapons *in which* Iran is superior is artillery, and this superiority exists largely in obsolete towed artillery weapons that have defensive value, but which are extremely difficult to use effectively in maneuver warfare.

Iran has developed a carefully focused capability to threaten shipping in the Gulf. This capability includes the purchase of three Russian submarines with minelaying capabilities, advanced naval mines, *and the* deployment of a wide range of anti-ship missiles on small craft and in land bases near the main shipping channels through the Gulf. It *also* includes the creation of a large force of Revolutionary Guards equipped for anti-ship and amphibious warfare. Iran has also focused its resources on obtaining long-range missiles and weapons of mass destruction, and the ability to fight unconventional warfare.

This kind of "focused poverty" allows Iran to get the maximum amount of regional influence and intimidation per Rial, although it has scarcely given Iran much war fighting capability against any regional coalition that involves the U.S.. Iran's efforts do, however, have major strategic *implications*. It has enough naval capability *stationed* along the Gulf coast, in the Strait of Hormuz, and deployable in the Gulf of Oman to harass shipping and require a major U.S. response if Iran should take offensive action.

At the same time, Iran is highly dependent on its oil export revenues and has no way to export any significant volume of oil except through the Gulf. It cannot defend any of its oil facilities against U.S. missile and stealth attacks, and its naval and anti-ship missile forces cannot survive for more than a few days to two weeks in the face of U.S. military action. Iran's mine warfare capabilities pose more of a threat in terms of long-term harassment, but they cannot block the Gulf. Iran lacks modern land-based air defenses, has limited modern fighter strength, has only about 30 modern attack aircraft (the Su-24), and has no modern airborne sensors and command and control assets. Its military forces and bases are open to U.S. retaliation.

Terrorism, Unconventional Warfare, Israel, and Lebanon

Iran has begun to debate its attitudes towards Israel since President Khatami's election. While their statements are often ambiguous and sometimes contradictory, both President Khatami and his foreign minister have indicated on several *occasions* that any peace *that* Syria agrees to will be seen as legitimate, that Iran's support of the Hezbollah only extends to the liberation of Lebanon, and that Iran will accept a peace that the Palestinian Authority agrees to. However, the Ayatollah Khamenei and Iran's traditionalists still see Israel as an illegitimate entity, and Iran's hard-liners see it as an enemy that must be destroyed. Unfortunately, it is the hard-liners that seem to control Iran's military liaison with Syria, *its* support of the Hezbollah in Lebanon, its acquisition of long-range missiles and weapons of mass destruction, and *its* support of anti-peace factions among the Palestinians like Hamas and Islamic Jihad. There is also broad concern *across factional lines* in Iran about Israel's long-range strike capabilities and nuclear weapons, and Iraq's continuing efforts to proliferate.

Israel is a long way away from Iran, and there is little *chance* of real conflict between the two countries. At present, however, Israel makes a convenient political whipping boy for Iran, and one where the whip is sometimes wielded out of real conviction. Until the peace process is revived, and receives broad Arab and Islamic support, Iran will probably continue to attack Israel's legitimacy and support Arab hard-liners. This may not involve the support of terrorist and extremist groups, or cooperation with Syria, but ideological differences are likely to combine with strategic interest, and the most the West may be able to hope for is *that* Iran *will* soften its rhetoric and not actively support violence.

In the interim, Iran has reduced its subversive and terrorist activity in the Gulf and Europe, but it has not reduced its shipments to the Hezbollah. Iran supplied the Hezbollah with new, longer-range rockets, although these may have been shipped before the election *of the moderate Khatami*. Mohammed Sadr, Iran's new Deputy Foreign Minister, first visited Damascus on September 9, 1997 to discuss the security situation in Lebanon and pledge continued military aid to the Hezbollah. Since then, Iran has continued to ship two plane-loads of arms a month to the Hezbollah, and Iranian senior officials have continued to meet regularly with Syrian officials and the leadership of the Hezbollah. Iran also continues to provide funds and paramilitary training to extremist Palestinian groups like the armed wing of Hamas and the Palestinian Islamic Jihad.

While Iran has noticeably improved its relations with Gulf states like Bahrain and Saudi Arabia, it has continued its intelligence surveillance of U.S. facilities like those in Saudi Arabia. It also bombed the bases of the People's Mujahideen, a violent opposition group based in Iraq, on September 29, 1997, and again in 1998 and 1999.

Weapons of Mass Destruction

Iran's ability to threaten shipping and tanker traffic in the Gulf represents the most serious problem that Iran *currently* presents in terms of energy security. In the medium to long-term, however, the problem will be Iran's efforts to produce biological, chemical and nuclear weapons, and long-range missiles. U.S. intelligence has strong indications that Iran is developing a Shihab long-range missile with ranges of 1,300 to over 2,000 kilometers with technical help from Russia, China, and North Korea. It also has strong evidence that Iran continues to develop chemical and biological weapons, and is importing equipment that is probably intended to develop fissionable material and nuclear weapons.

At the same time, the evidence relating to the speed and scale of Iran's nuclear weapons program is ambiguous. Khatami replaced Reza Amrollahi, the head of the Atomic Energy Organization of Iran, with Gholamreza Aghazadeh, Iran's former oil minister. The reasons for this appointment are not clear. Some sources argued that it represented an effort to improve the administration of Iran's nuclear programs (Amrollahi had developed a reputation as an awful administrator and manager). Some felt it might be part of an effort to make Iran's nuclear power program more efficient, while others saw it as part of an effort to review whether such a program was cost-effective at all. *Still others felt* it may represent a down-playing of Iran's nuclear weapons program.

Aghazadeh did, however, reaffirm Iran's commitment to a massive nuclear power program on October 3, 1997. At a meeting with Hans Blix, the head of the International Atomic Energy Agency (IAEA), Aghazadeh indicated that Iran planned to add a second 1,000-megawatt generating unit to its existing efforts to build a 1,000-megawatt unit in Bushehr, and eventually produce 20% of Iran's electric power needs from nuclear units. He indicated that Iran had approached Russia to buy two more 440 megawatt reactors and was seeking an eventual total of six, and that it was still seeking two 300 megawatt nuclear reactors from China.

Iran also revised its arrangements with Russia in 1998 to increase the Russian share in the construction and support activities at Bushehr, and did so in spite of a major drop in Iranian oil revenues. While some internal debate has taken place over the extent to which Bushehr has become an expensive white elephant, Iran remains committed to the project at this high level.

Iran stepped up its long-range missile program in 1998 in spite of its revenue program and tested its first indigenously produced long-range missile, the Shihab 3. It continues to import or try to import, key technologies for biological and nuclear weapons. Although senior U.S. officials stated that Iran was probably 5-10 years away from acquiring nuclear weapons in June 1999, Iran will probably acquire extremely lethal biological weapons much sooner, and could assemble nuclear weapons much more quickly if it could assemble fissile material from an outside source

Iran and the Geopolitics of Energy

Whatever happens, Iran will have vast strategic importance in shaping the future of the world's energy balances. The U.S. Department of Energy estimates that Iran holds 93 billion barrels of proven oil reserves, or roughly 9% of world's total. The vast majority of Iran's crude oil reserves are located in giant onshore fields in the Khuzestan region near the Iraqi border and Persian Gulf terminus. More than half of Iran's 40 producing fields contain over one billion barrels of oil. The onshore Ahwaz, Marun, Gachsaran, Agha Jari, and Bibi Hakimeh fields alone account for about two-thirds of Iran's oil production. Most of Iran's crude oil is low in sulfur and light, with gravities in the 30o-39o API range. Iran has not had significant oil exploration activity in nearly 3 decades.

Iran is OPEC's second largest oil producer, with an average current output of *3.47* million bbl/d (MMBD), nearly all of which is crude oil. This compares to a current (April 99) OPEC crude oil production quota of *3.359* MMBD. Iran's current sustainable production capacity is estimated at around

4 MMBD, but this figure is controversial, with some claiming that Iran has maintained production levels at some older fields only by using methods that have permanently damaged the fields.

The U.S. Department of Energy estimates that Iran will expand its oil production capacity from an average of 3.9 million barrels per day in 1997, to 4.3 MMBD in 2005, 4.5 MMBD in 2010, and 5.5 million barrels per day in 2020. To put these numbers in perspective, the Department of Energy estimates that Iran *will maintain a steady 5% share of world production from the present to 2020.*

Iran is also a major gas producer. Iran's strategic position has a major impact on the development of energy resources in the Caspian and Central Asia. *It contends* that treaties signed in 1921 and 1940 are still valid, implying that all countries bordering the Caspian must approve any offshore oil developments. In late February 1998, Iran's Foreign Minister Kamal Kharrazi reiterated Iran's position that any unilateral exploitation of Caspian Sea resources would be illegal. Oil Minister Zanganeh has stated that Iran backs national zones extending several miles from the coast and a "condominium" in the middle of the Sea. Iran also has stated (along with Russia) that it opposes laying an oil pipeline across the Caspian Sea floor.

Iran sees itself as a natural transit route for oil and gas exports from the landlocked Central Asian countries to world markets. This vision is complicated, however, by political considerations, particularly the U.S. policy of opposing pipelines through Iran, the shortest (and most likely the least expensive) path to the open sea. As part of its attempt to isolate Iran and to contain its influence in the region, the United States has instead favored multiple routes for Caspian oil and gas through the Caucasus region to the Black sea or to the Turkish port of Ceyhan.

Iraq

Iraq's is one of the most troubled and repressive states in the world. It has vast oil resources and great potential wealth, but it is a nation that has been in an almost continuous state of crisis since the late 1970s. There are few prospects that things will change decisively as long as it is under its present regime, and the aftermath of the Gulf War has scarcely improved this *situation.* For nearly a decade, the "war of sanctions" has kept Iraq under of mix of sanctions, inspection regimes, and export and import controls that have *left it politically* isolated and *economically* crippled. During this time,

Iraq has been deprived of overt access to arms imports and the technology it needs to proliferate. Iraq's neighbors, including Iran, have been able to continue their efforts to proliferate and build-up their conventional forces, as well as develop their economies.

Kuwait remains independent, but Iraq has been forced to transfer key territory near its small coastline on the Gulf to Kuwait. Until December 1998, Iraq was also forced to allow the UN Special Commission (UNSCOM) and International Atomic Energy Agency (IAEA) to supervise the destruction of most of its weapons of mass destruction and ability to produce them, *and* to allow Coalition forces to enforce military restrictions and 'no fly zones.' It has also been forced to tolerate Coalition efforts that have created an independent Kurdish enclave in Northern Iraq.

The 'War of Sanctions'

Saddam Hussein has repeatedly shown that he has three major priorities: his own survival, the rebuilding of his conventional military forces, and the preservation of his capability to manufacture and deploy weapons of mass destruction. From the first days of the cease-fire to present, he has systematically attempted to violate the terms of the cease-fire. He has fought and won a brutal civil war against his Shi'ite opposition in the south, and kept up constant pressure on the Kurdish enclave in the north. He has mobilized and deployed his army towards Kuwait. He has constantly challenged the UN's efforts to destroy his weapons of mass destruction.

Since the cease-fire in early 1991, there has never been a three-month in which Saddam Hussein has not provoked a new confrontation with the UN, his neighbors, or the West. He has systematically impoverished his people, and mortgaged their hopes for future economic development by concentrating Iraq's scarce resources on rebuilding his military forces. He has refused economic aid and relief from limits on Iraq's capability to export oil in an effort to break out of sanctions. He has made constant efforts to divide the Arab world, and has courted key nations like France and Russia with oil deals and promises of future economic concessions. To all practical purposes, he has turned his defeat and the cease-fire into an extension of war by other means.

The sanctions crisis that Saddam Hussein provoked beginning in the fall of 1998, and which led to the suspension of UNSCOM and IAEA inspection efforts in December 1998, is simply another step in Iraq's struggle to break out of sanctions without meeting the terms of the UN cease-fire. *There has been some new challenge to the UN every fall since the end of the Gulf War.*

At the same time, Iraq has steadily intensified its efforts to exploit "sanctions fatigue" and has found that it can use UNSCOM as a tool to divide the Security Council. Iraq is increasingly coupling its efforts to rebuild its military forces with propaganda that exploits the hardships of the Iraqi people, the near-collapse of the Arab-Israeli peace process, the concern Arab nations feel about Iraq's sovereignty and territorial integrity, and the fear many Southern Gulf states still have of Iran. The individual battles in the "war of sanctions" have often focused on military and security issues, but the strategic goals behind the war are much broader and have clear links to Iraq's "strategic culture.

Proliferation and the Continuing Iraqi Military Threat

This mix of history and current strategic priorities helps to explain why Saddam Hussein has continued to commit Iraq to new military confrontations with the U.S.-led Coalition. It explains why Iraq was willing to forgo over $120 billion in oil export revenues during 1991-1996, and why Iraq's present leaders will continue to use every possible means to break out of UN sanctions import restrictions. At least for the foreseeable future, Iraq's official policy towards the rebuilding of its conventional forces and proliferation will continue to be a mix of denial and lies, with occasional bluster and indirect threats. These lies will *be* told to Iraq's people, media, intellectuals, and military officers, as well as to other nations.

Iran may be the rising military power in the Gulf, but Iraq's conventional military forces have been extensively reorganized since the Gulf War, and have regained a substantial part of their pre-war military capabilities. Iraq can still deploy massive land forces against Kuwait and the Eastern Province of Saudi Arabia, and Iraq's conventional forces remain the largest in the Gulf in many areas of conventional force strength.

While Iraq has lost many of the capabilities it possessed at the time of the Gulf War, its half-decade long struggle to preserve its capability to produce and deploy weapons of mass destruction and long-range missiles continues. It is clear from United Nations reports that Iraq retained a significant capability to deliver chemical and biological weapons before UNSCOM was forced to leave Iraq in December 1998, and had some of the components necessary to rebuild a long-range missile force. It is equally clear that Iraq conducted a covert effort to rebuild its capabilities and to rapidly expand them the moment UN sanctions weaken, even when the IAEA and UNSCOM were active in Iraq.

These Iraqi military capabilities take on a special meaning because Saddam Hussein and his coterie have repeatedly demonstrated that they are willing to take extreme political and military risks with little warning. Iraq's near-genocidal attacks on its Kurds, and decision to use chemical weapons against Iran, are examples of its willingness to take such risks and ignore world opinion. Iraq's attack on Iran, its invasion of Kuwait, and its *sudden* missile strikes, *are* secret shifts in policy *made* by a small decision-making elite, possibly *even* one man. In each case, the warning indicators were ambiguous and many regional leaders and experts argued that Iraq would take a much more moderate course of action.

It is equally dangerous to try to predict the extent to which Iraq will escalate a crisis once it begins. The scope of Iraqi military action expanded sharply during the course of its war with its Kurds, the Iran-Iraq War, and invasion of Kuwait. Iraq's leaders have not been indifferent to threats to their own survival, but they have often proven willing to escalate in ways that neither their neighbors nor Western experts predicted.

Iraq must be regarded as a major military threat to the security of the world's supply of oil exports. There is little *hope* that Kuwait can be safe as long as any leader like Saddam Hussein is in power, unless the U.S., its Gulf allies, and other Coalition powers maintain a strong deterrent and warfighting capability to deal with the Iraqi threat. There is a continuing risk of a further conflict between Iraq and Iran, although no one can dismiss the possibility of some alliance of convenience between the two regimes. The Kurds remain a major issue, as does the instability along the Iraqi-Turkish border. Saudi Arabia has a long and vulnerable border with Iraq, and has done far too little since the Gulf War to improve the defense of its oil-rich Eastern province. Iraq remains a potential threat to Israel and Jordan, and the Arab-Israeli peace process.

Iraq will make every effort to conceal its true plans and the full nature of its military efforts, and only Saddam Hussein and a few trusted supporters will have any overview of Iraq's military progress and capabilities. Furthermore, Iraq's plans and polices will remain opportunistic and *erratic*. Iraq's leaders will be unable *to* predict the exact areas where they will be successful in evading or vitiating UN sanctions and controls. As a result their strategy, military doctrine, and force development efforts can be expected to evolve on *a basis of* opportunity. The only thing that seems certain is that Iraq will make a continuing effort to obtain advanced conventional arms and to proliferate in every way that Iraq can conceal.

Importance of Iraqi Oil Exports

At the same time, Iraq is critical to meeting future world oil demand. The U.S. Department of Energy estimates that Iraq will increase its oil production capacity from 2.2 million barrels per day in 1990, and 1.6 million barrels per day in 1997, to 2.8 (2.3-2.9) million barrels per day in 2000, and then to 5.9 (4.7-7.2) million barrels per day in 2020. To put these numbers in perspective, the Department of Energy estimates that Iraq will increase from less than 2% of world production in 1997, to 3.5% in 2000, 4% in 2010, and 5.2% in 2020.

Kuwait

Kuwait plays a critical role in any projection of the world's future oil supplies. Kuwait contains an estimated 96.5 billion barrels of proven oil reserves, over 9% of the world total. Before recent cutbacks in production because of low oil prices, Kuwait was producing an estimated 2.4 million barrels per day (bbl/d), including 250,000 bbl/d of Neutral Zone production. Kuwait is one of the two major oil exporting powers that can rapidly increase production in an emergency, and its role as a swing producer may increase over time. The U.S. Department of Energy estimates that Kuwait will expand its production from 2.6 million barrels per day in 1997, to 3.2 million barrels per day in 2010, and 5.2 million barrels in 2020.

Kuwait's Security

Kuwait's national Security is a paramount issue. Kuwait's geography, small size, and limited population make it one of the most vulnerable Gulf states. Its location on Iraq's border has been the source of continuing Iraqi threats, military confrontation, and actual invasion.

Kuwait, however, will pose a continuing challenge to U.S. power projection, particularly as Iraq rearms and rebuilds its weapons of mass destruction. Kuwait is located in the far northwestern corner of the upper Gulf between Iraq and Saudi Arabia, and within a short distance of Iran. It is one of the world's major oil powers, but it has a total area of only 17,800 square kilometers — roughly the size of New Jersey. At its largest point, Kuwait is about 200 kilometers from north to south and 170 kilometers from east to west. It shares a 242-kilometer border with Iraq and a 222-kilometer border with Saudi Arabia. It has a 499-kilometer coastline on the Gulf, and its territory includes nine islands. Bubiyan and Warbah — two large islands in the north — are uninhabited but are of strategic importance, because they border the Umm Qasr channel, which is Iraq's only waterway with direct access to the Gulf.

Kuwait, the U.S., and other allied nations have already been forced to react to Iraqi provocations in 1992, 1994, 1996, and 1997. Even if these did not affect Kuwait's overall security, *they posed major threats* in the northern border area. As part of the cease-fire terms, the UN set up a special UN Iraq/Kuwait Boundary Demarcation Commission. As a result, correcting the border to the original line moved it to the north and into territory that Iraq had occupied before World War II.

Kuwaiti Oil and the Continuing Threat from Iraq

The new border offered Kuwait considerable advantages at the expense of Iraq. It gave Kuwait greater control over the Ratga and Rumalia oil fields in its northern border area, and reduced Iraqi access to the port facilities at Umm Qasr. At the same time, the new border created political problems. Only six days after the Secretary-General accepted the report, the Speaker of Iraq's National Assembly stated that the new border would keep tensions in the region high. Iraq refused to accept the new demarcation and Iraqi editorials, and Iran's media made new claims to Kuwait as Iraq's 19th province.

Iraq's leadership is unlikely to end its claims to part of Kuwait's Ratqa oilfield even if it accepts the existence of an independent Kuwait. This field was once thought to be an independent reservoir, but is actually a southern extension of Iraq's super-giant Rumaila field. Making matters worse, Iraq disputed its border with Kuwait, and in fact claimed the entire country as part of Iraqi territory. During the weeks preceding Iraq's August 1990 invasion of Kuwait, Iraq accused Kuwait of stealing "billions of dollars" worth of Rumaila oil, and had refused to negotiate a sharing or joint development arrangement for Ratqa and southern Rumaila. Kuwait's new U.N.-drawn border includes a 1919-foot extension for Ratqa further to the north.

It seems likely that Iraq will create new problems in the border area the moment it is given the political and military opportunity. Kuwait has already had to delay plans to allow Western oil companies to explore and develop its oil fields near the Iraqi border because of the risk of new clashes. Iraq, on the other hand, is aggressively attempting to negotiate deals with nations like Russia to exploit the fields on its side of the border once the UN sanctions are lifted.

Pivotal Role of U.S. Power Projection Forces

Kuwait is also a symbol of the critical role that U.S. power projection plays in ensuring the security of Gulf energy exports. As early as August and September of 1992, the confrontation between Iraq and the UN over the elimination of weapons of mass destruction, and Iraq's treatment of its Shi'ites and Kurds, forced Kuwait and the U.S. to transform their joint exercises into a demonstration that the U.S. could protect Kuwait against any military adventures by Iraq.

The U.S. rushed Patriot batteries to both Kuwait and Bahrain, and conducted a test pre-positioning exercise called Native Fury 92 and an amphibious reinforcement exercise called Eager Mace 92. The U.S. also deployed a 1,300 man battalion from the 1st Cavalry division, 1,900 Marines and 2,400 soldiers, including two armored and two mechanized companies.

Since that time, the U.S. has repeatedly had to use force to contain Iraq, maintain a near-permanent carrier task force in the Gulf, keep a brigade set prepositioned in Kuwait, base aircraft in Turkey and the Southern Gulf to maintain the 'no fly' zones in Iraq, and maintain a network of force deployments and prepositioning capabilities in virtually every Southern Gulf country. The U.S. and Britain launched cruise missile strikes on Iraq on two other occasions, and conducted major air and missile strikes on Iraq in December 1998, as part of Operation Desert Fox. Since that time, the U.S. and Britain have fought a steady low-level air defense war with Iraq.

If U.S. and British forces were not deployed in the Gulf, it is doubtful that Kuwait would have survived Iraq's desire for revenge, and it is uncertain that the U.S. would have *been able to maintain sanctions or intervene* a second time. Similarly, U.S. forces act as a powerful deterrent to Iranian efforts to use its ability to threaten maritime traffic through the Gulf and Strait of Hormuz. They also act as a major counterweight to Iranian and Iraqi proliferation, and help give meaning to U.S. efforts to limit arms transfers to Iran and Iraq.It is uncertain how long U.S. forces will or can be deployed forward at their current strength, and it seems likely that any combination of the fall of Saddam Hussein and sustained moderation in Iran might lead to the withdrawal of most U.S. forces move out of the Gulf and over-the-horizon. Even then, however, the U.S. will still play a critical role in securing the 'four cornered' balance of power in the Gulf discussed earlier. Neither Saudi Arabia nor any combination of Southern Gulf states seems likely to be able to create a significant deterrent and defense capability to deal with either Iran or Iraq in the near to mid-term. Proliferation makes this equally true in the long-term. The U.S. may not need to be forward and visible, but unless it is capable and credible, the fragile structure of military security in the Gulf *could* explode at any time.

The United Arab Emirates (UAE)

The UAE bears a certain resemblance to the Holy Roman Empire in one respect: It is not united, most of the population is not Arab, and its components may not stay emirates. It is a rentier state of seven small city-states or emirates. Most of its population consists of foreign workers, and there are long-standing tensions between its largest Emirates (Abu Dhabi, Dubai, and Sharjah), and the less wealthy mini-states to the east.

Uncertain Unity, Mostly Non-Arab, but Oil-Rich and Emirate Oil and gas wealth, and vulnerability, are the glue that holds the UAE together. *Despite its disunity, the UAE is on the world's largest oil producers.* The U.S. Department of Energy estimates that the UAE will increase its production capacity from 2.5 million barrels per day in 1990, and 2.7 million barrels per day in 1997, to 2.8 (2.6-3.0) million barrels per day in 2000, 3.2 (3.0-3.3) million barrels per day in 2005, 3.4 (3.1-4.0) million barrels per day in 2010, 4.2 (3.3-4.7) million barrels per day in 2015, and 4.9 (3.7-5.7) million barrels per day in 2020.

Much has also depended, however, on the political skills of Sheik Zayed, the ruler of Abu Dhabi, and his ability to use diplomacy and oil wealth to hold the UAE together. Sheik Zayed is aging and his sons have yet to show they can replace his political skill. Dubai is particularly conscious of its status, and there are low-level concerns that the Eastern Emirates might seek independence or that Oman might seek to include them in its territory. Political reform is moving very slowly in the UAE. It is still wealthy enough to coopt most potential opposition, but it also is creating a new class of intellectuals who will probably seek participation in government and who are serious critics of many aspects of U.S. policy. Key issues are the presence of U.S. power projection forces, the Arab-Israeli peace process, arms sales to the UAE, and the "hardship" sanctions have imposed on the Iraqi people. Many intellectuals are now calling for a Gulf security structure that eliminates a U.S. presence.

As yet, there are few signs of internal political tension. There have, however, been serious riots between Hindu and Pakistani foreign workers. The UAE has a total population of about 2.30 million and 76% is expatriate, with 30% Indian, 16% Pakistani, 12% Arab, 12% other Asian, and 1% European. The military is at least 30% expatriate. There is at least some question of how long the landlord will actually control the country.

Potential Foreign Threats: The Problem of Iran

The UAE's principal foreign threat is Iran, which is also a major trading partner – particularly with Dubai. The the key to this dispute *between* the two countries is the sovereignty of three islands that divide the major shipping channels in the Gulf: Abu Musa, the Greater Tunbs, and Lesser Tunbs. Iran effectively seized all three islands from the nominal control of Ras al Khaimah when the British left the Gulf in 1971, and did so with at least tacit British compliance. Ras al Khaimah, which later became part of the UAE was given a face-saving compromise and jurisdiction over the small Arab population on Abu Musa. The island dispute is further complicated by the development of Mubarak field. It is located six miles off Abu Musa, and has been producing gas-rich oil since 1974.

This strange combination of de facto Iranian sovereignty and a limited political role for the UAE survived the Iranian Revolution and the Iran-Iraq War. It came as something of a surprise, therefore, when Iranian troops effectively occupied all three islands in 1992. In 1995, the Iranian Foreign Ministry went further, and claimed that the islands were "an inseparable part of Iran. " Iran rejected a 1996 proposal by the Gulf Cooperation Council (GCC) for the dispute to be resolved by the International Court of Justice, an option supported by the UAE. In early 1996, Iran took further *steps* to strengthen its hold on the disputed islands. These actions included starting up a power plant on Greater Tunb, opening an airport on Abu Musa, and announcing plans for construction of a new port on Abu Musa.

The UAE has received strong support from the GCC, the United Nations, and the United States. The UAE officials in Abu Dhabi have chosen not to actively escalate the territorial dispute, in part because Iran is one of Dubai's major trading partners, accounting for 20% to 30% of Dubai's business. The UAE has, however, repeatedly reasserted its claims to sovereignty, sought support from bodies like the GCC, and tried to bring the issue to the World Court. It seems doubtful that the dispute will escalate to open conflict, particularly since President Khatami has shown some flexibility over the issue. Nevertheless, the dispute is a key *reason* for the UAE's military build-up.

Though little progress has been made between the UAE and Iran, the dispute over the islands did create some animosity between the UAE and Saudi Arabia. Throughout 1999, Saudi Arabia and Iran made attempts to reconcile their differences and improve their relations. These attempts were met with harsh criticism from the UAE who were angered by what they felt was Saudi abandonment of their claim to the islands. The two sides exchanged heated remarks, and the dispute seemed to have the potential to become a serious issue between the UAE and Saudi Arabia, but it was resolved through Qatari mediation.

Other potential disputes include past Omani claims to territory in the UAE, and past border problems with Qatar and Saudi Arabia.

Other Southern Gulf States

Bahrain, Qatar, and Oman are not major oil producers, but any geopolitical analysis of the Gulf must consider the risk that the rise of a radical or unfriendly regime could pose in any of these states. Qatar is also a major potential gas producer, with the third largest reserves in the world. All three states play an important role in U.S. power projection. Bahrain is the host to the U.S. 5th Fleet, Qatar has agreed to allow the U.S. to preposition a brigade set on its soil, and Oman provides prepositioning and port facilities.

Bahrain

The principal risk that Bahrain presents is that the long-standing political tensions between its Sunni elite and Shi'ite majority could explode into open civil conflict, lead to Iranian covert or overt intervention, and/or bring down its royal family. These problems are compounded by serious structural economic problems that are the result of the depletion of oil reserves, a growing population, over-dependence on foreign labor, and the over-concentration of wealth in the hands of a relatively small elite. Bahrain also has long-standing tensions with Qatar relating to a feud between the two royal families, and rival claims to a chain of small islands between them.

Oman and the Strait of Hormuz

Oman's oil and gas exports play a role in future global energy balances. Oman has proven oil reserves of 5.3 billion barrels, and natural gas reserves of 28.4 trillion cubic feet. However, its primary strategic importance lies in its control of the Strait, the strategic position of the Omani enclave in the Musandam Peninsula opposite Iran, and its value in staging and prepositioning U.S. power projection forces. The Strait of Hormuz is the only shipping channel in and out of the Gulf. Over 14 million barrels per day (b/d) of oil flow through this Strait to Japan, the United States, Western Europe, and other countries. This makes the Strait the world's most important oil chokepoint. At its narrowest, it consists of 2-mile wide channels for inbound and outbound tanker *traffic* within the Omani side of the Strait, and a 2-mile wide buffer zone.

Even partial closure of the Strait of Hormuz would *force* tankers to use Red Sea ports and the use of longer alternate shipping routes at increased transportation costs. These alternative routes include the 4.8 million b/d capacity Petroline, the 2.2 million b/d IPSA 1 and 2 lines, and the Abqaiq-Yanbu natural gas liquids line across Saudi Arabia to the Red Sea. *There are no alternate routes that can meet anything approaching current export levels, much less the much higher production levels forecast by DOE, if any major interruption took place in the flow of oil through the Strait.* To put these projections in perspective, the DOE reference case implies that up to three times *as many* tankers will transit the Strait in 2020 than today. This means that Oman will become steadily more important with time, *especially since* Oman also offers the only potential secure land route for pipelines from the Southern Gulf to ports on the Indian Ocean.

Oman faces a number of security problems. Its most tangible foreign threat is Iran. Although Oman has made continuing efforts to improve its relations with Iran, and has had considerable success in dealing with Iran under both President Rafsanjani and President Khatami, it sizes and shapes its forces primarily to deal with the Iranian threat. Oman is fully aware of the Iranian build-up in the lower Gulf and the nature of Iranian proliferation. It is also aware that Iran may seek to *seize* control of the Strait of Hormuz in some future conflict, although the main shipping channels flow through the Omani side.Oman faces lesser threats *due to* long-standing tensions with Saudi Arabia. Oman and Saudi Arabia have formally demarcated their boundaries, but it is not clear that Saudi Arabia has firmly given up all of its past claims to Omani territory. The UAE has expressed some fears that Oman might seize its eastern emirates, and Oman and Yemen *have* had border clashes in the past. None of these threats seem particularly important today, but the Gulf is a region where no problem ever seems to have a final solution or to be forgotten.

Oman also has many of the same internal problems that Saudi Arabia faces, but with far less energy reserves and wealth. Sultan Qabus has no clear successor. It has significant oil and gas riches, but it is anything but "oil rich' by the standards of its neighbors, and it has been hit hard by the current 'oil crash.' Oman's current five-year plan calls for unrealistic levels of foreign investment, and future gas and oil production. There is some salination of key agricultural areas, and industrial development has been slow.

Oman's development program and stability have been hit hard by low average energy prices and the "oil crash" of 1997-1998. They are threatened by an extremely high population growth rate *and* a declining real per capita income in spite of major advances in overall national development. Oman also has failed to reduce its dependence on foreign labor at anything like the rate required to ensure employment and productive jobs for native labor. It is far too soon to predict a serious threat of political instability, but Oman's current plans outreach its grasp and its government relies more on hope than action.

Qatar

Qatar is a small nation whose importance lies largely in its gas resources and the fact that it is a peninsula in a strategic location in the middle of the Southern Gulf than any other factor. Qatar is also one of the founders of the Gulf Cooperation Council (GCC), and Qatari troops fought on the side of the UN Coalition during the Gulf War. The Qatari forces at the battle of Ras al Khafji were among the first Coalition troops to engage Iraqi ground forces. Since the Gulf War, Qatar has steadily expanded its strategic ties to the West and the United States. It is providing prepositioning facilities for a U.S. Army brigade and the support equipment for a division base, and conducts regular exercises with British, French, and U.S. forces.

Qatar is still a significant oil exporter, although its reserves are limited compared to those of many of its Gulf neighbors. Qatar has proven oil reserves of 3.3-4.3 billion barrels (0.4% of the world's total). The U.S. Department of Energy estimates that Qatar's oil production capacity will increase slowly from 0.6 million barrels per day in 1997, to 0.7 million barrels per day in 2020. Qatar's main role in global energy balances is, however, that it has massive gas reserves and is becoming a major gas power. According to U.S. estimates, Qatar has total producible gas reserves of approximately 237 trillion cubic feet, or 40 billion barrels of oil equivalent. This is about 5% of the entire world's gas reserves and Qatar is the third largest nation in the world in terms of total gas reserves. Qatari sources make much more ambitious claims and estimates that Qatar has reserves approaching 380 trillion cubic feet.

Qatar faces a potential challenge from Iran because its massive North Field has gas reserves that cross into Iranian waters. Qatar has good relations with Iran, however, and the major risk that could affect its near to mid-term stability is internal politics.

The present Amir, Sheik Hamad bin Khalifa Al Thani, assumed power in June, 1995 as the result of the kind of "family coup" that has occurred several times in Qatar since the Al Thani family first took power. He has already survived one low-level coup attempt, and seems reasonably secure although his health is uncertain. There are no current signs that there would be any radical changes in Qatar's policies, even if a new Amir does emerge from within the royal family.

Political stability is a different issue. The succession within the Al-Thani family has *been* a complex series of power struggles between and within branches of the ruling family. Under the Basic Law of 1970, the Amir must be chosen from and by the adult males of the Al Thani family, but there are no formal procedures for determining who shall rule. The exact powers that can be assumed by a given Amir vary according to ruler, although custom establishes a number of practical limits on the ruler's power.

Qatar also has a tendency to feud with its Southern Gulf neighbors and other moderate Arab states. The Foreign Affairs Minister, Sheik Hamad bin Jassem bin Jabr Al-Thani, has sometimes seemed to delight in such feuds. He has picked them with Bahrain, Egypt, Saudi Arabia, and the UAE – although Qatar's feud with Bahrain is the only one which is currently active.

Qatar is another symbol of the fact that there will be no cohesive security structure in the Gulf and that only U.S. power projection capabilities can provide the balance necessary to ensure some degree of stability.

Energy Politics and the North African States

The North African states are relatively small oil exporters by the standards of the Gulf, but they are still very important to the world market. Total regional oil exports were 2.3 MMBD in 1995, and the DOE projects that this total will only increase to 2.6 MMBD in 2020. As a result, North African oil exports will drop from 6% of world exports in 1995 to around 4% in 2020. This is still an important part of the world market, *however*, and will far exceed the estimated exports of Asia (0.3 MMBD). West Africa, however, is estimated to produce 3.1 MMBD, the North Sea to produce 4.1 MMBD, South America to produce 4.3 MMBD, and FSU to produce 5.6 MMBD.

North Africa serves key markets in Southern Europe, and Algeria also has 131 trillion cubic feet *in* gas *reserves*. While this is only about 2.6% of world gas reserves, it means that North Africa is also an important gas exporter to Europe. The fact that Morocco, Algeria, Libya, and Tunisia make up a good portion of the southern coast of the Mediterranean give them added importance to Europe, as does the growing problem posed by emigration from these countries to nations like France, Spain, and Italy. The end of the Cold War has largely ended the regional arms race and much of the terrorist threat from Libya, but many Europeans see internal instability in North Africa as posing a much more tangible demographic threat than the military threats of the Cold War.

Algeria

Algeria continues to be the scene of a low-level civil war between its corrupt military junta and extremists who ruthlessly murder innocent civilians in the name of Islam. At the same time, it has failed to implement serious economic reform, faces growing tensions between Arabs and Berbers, and has disastrous demographics. As much as 40% of its young males have no job or a meaningless state subsidized job with token wages.

The Problem of Civil War The death toll from the Algerian government's *seven*-year conflict with the *Islamic Salvation Front (FIS)* and the Armed Islamic Group (GIA) was estimated at over 100,000 by the middle of 1999. Much of the fighting has been centered in Algiers and coastal towns in the country's northern area. The conflict started after the government's annulment of *the* results from the December 1991 general elections. Escalating political violence and terrorism then led to government repression and the imposition of a state of emergency. While the level of violence has diminished since 1997, and the FIS is now pursuing more peaceful options, it has scarcely ended.

Algeria has taken some steps towards reform and democracy. Parliamentary elections were held in early June 1997. Out of 380 seats in the new legislature, the newly formed National Democratic Rally (RND from French acronym) won 155 seats and the National Liberation Front (FLN) won 64 seats. Two moderate Islamist parties, Movement of Society for Peace (formerly Hamas) and Ennahdha, took 69 seats and 34 seats respectively. Opposition parties, represented mainly by Front des Forces Socialistes (FFS) and Rally for Culture and Democracy (RCD), each won 19 seats. Remaining seats were apportioned between various other small parties. Following the elections, opposition parties complained bitterly about voting irregularities and some UN observers indicated that they did not have free access to all polling stations. Despite the controversy, the Organization of African Unity and the Arab League of Nations gave their stamp of approval to the election results.

Local and regional elections were held on October 23, 1997 with similar results. The winners in these elections were the two pro-government parties, RND and FLN, and the Movement of Society for Peace. *Former* President Zeroual's RND party won more than half the seats at stake in the local races and scored similar victories in the regional races. These elections were marred by even more controversy than the parliamentary elections. Pro-government and opposition parties alike protested the results claiming widespread fraud and called for the results to be revised or thrown out. In the aftermath, 30,000 people took to the streets of Algiers on October 30, 1997 in protest, while several political parties tried to shut the country down by calling for a general strike on November 12, 1997. Workers heeded government warnings not to observe the three-hour work stoppage, and Algerian police cut off plans for a national march set for November 13, 1997.

High levels of violence and unrest continued in Algeria into 1998. While the fighting declined during the course of 1998, new problems emerged. There were growing reports that Algeria's military junta had come to view Zeroual as too compromising and too soft on Islamic extremists. On September 11, 1998, Zeroual suddenly announced that he would not complete his five-year term and would call for an early presidential election in February 1999. Zeroual later postponed the election until April 15, 1999 in order to allow the various political parties time to organize and prepare.

The campaign began with 48 potential candidates, leading many Algerians to hope that the presence of genuine opposition parties might signal the beginning of a shift in the military's attitude towards more political openness. However, only one candidate remained by the day of the election, former Foreign Minister Abdelaziz Bouteflika. Many of the former candidates complained that the military had pressured them and had stuffed ballots to ensure Bouteflika's election.

It is still unclear in what political direction Algeria is headed. Nevertheless, it seems most likely that there will only be slow changes in the status quo and repression. Bouteflika has moved towards peace, but he seems to lack the understanding of the need for market reforms in Algeria, and has expressed the need to curtail the freedom of the press. Unfortunately, it seems that Algeria's "democratization" will contune to remain rhetorical rather than actual.

Civil War, the "Oil Crash,' and Energy

Until the drop of oil prices in late 1997, Algeria was able to achieve overall economic growth in spite of these conflicts. The problem is that such growth could not compensate for the political turmoil and massive economic mismanagement that occurred in previous years, and population growth. Algeria's population has risen from 18.7 million in 1980 to 29.3 million. As a result, the per capita income has dropped from a peak of about $1,800 in 1985 to around $1,500. This is scarcely "development," and the decline in real wealth per capita interacts with rising expectations, social violence, and massive unemployment.

So far, civil war has had only a limited impact on Algeria's exports of oil and gas. Most of Algeria's political violence involves FIS and GIA attacks on the country's army and police force, public officials, and civilians, although some foreigners have been targeted. The country's oil and natural gas industry continues to remain relatively unscathed, due in part to the remote southern location of most oil and gas fields and production facilities.

In April 1995, the government created four "exclusion zones" to protect oil and gas facilities and personnel in the producing centers of El Oued, Laghouat, Illizi, and Ouargla. Within these zones, traffic and shipments are regulated by army and police units. Most oil and gas pipelines, though unguarded, are deeply buried and well protected against sabotage. However, the TransMediterranean pipeline, linking Algeria and Italy, experienced a brief disruption in November 1997. A fire caused damage to two valves located about 12 miles from the Tunisia border. The fire was reportedly set by anti-government forces, but a particular group has not been identified.

The security of these energy resources and their development is an important geopolitical concern. Algeria is considered to be under-explored. Algeria's National Council of Energy believes that the country still contains vast hydrocarbon potential and the government and its state-owned Sonatrach have put forth great efforts in recent years to change this situation. Significant oil and gas discoveries have been made over the last four years, *mostly* by foreign companies. Under a government program for 1996-2000 launched in April 1996, Sonatrach and its foreign partners will attempt to increase Algeria's crude oil production to 1.4 million bbl/d by 2000. Included in this program are provisions for 300 exploration wells to be drilled between 1996 and 2000, half by Sonatrach and the other half by foreign companies.

Algeria's Importance to World Energy Balances

Official estimates of Algeria's proven oil reserves still remain at 9.2 billion barrels, but they are expected to rise significantly as a result of recent oil discoveries, plans for more exploration drilling, improved data on existing fields, and use of enhanced oil recovery (EOR) systems. The U.S. Department of Energy estimates that Algeria will increase its oil production capacity from 1.4 million barrels per day in 1997, to 2.1 million barrels per day in 2015, and 2.0 (1.9-2.0) million barrels per day in 2020.

Algeria *will* increase its crude oil exports over the next few years due to a rapid shift towards domestic natural gas consumption and planned increases in oil production by Sonatrach and its foreign partners. Approximately 90% of Algeria's crude oil exports go to Western Europe, with Italy as the main market followed by Germany and France. The Netherlands, Spain and Britain are other important European markets. The United States makes up a significant portion of the remaining 10% of Algerian crude exports. Algeria's Saharan Blend oil, 45° API with 0.05% sulfur and negligible metal content, is among the best in the world.

Algeria contains 130 trillion cubic feet (Tcf) of proven natural gas reserves, ranking it in the top ten worldwide. Sonatrach estimates that Algeria's ultimate gas potential is around 204 Tcf, of which 135.5 Tcf is recoverable. Algeria's largest gas field is the super-giant Hassi R'Mel, which initially held probable and possible reserves of between 95-105 Tcf and proven reserves of about 85 Tcf.

The Department of Energy reports that Algeria hopes to boost its natural gas exports from 3.8 billion Bcf/d at present to 5.8 Bcf/d by 2000. Through 2000, Italy and France are expected to remain the largest purchasers of Algerian gas, with 47 percent and 20 percent shares respectively. Through 2020, the European Commission has forecast that Algerian exports will not exceed a 25 percent share of the European gas market. In contrast, Europe's two other major gas suppliers, Norway and Russia, each are expected to maintain or expand their 25 percent market shares. In addition, Algeria has had a long-standing policy to develop its gas reserves as a source of domestic energy and as a raw material for the petrochemical industry. As of mid-1997, approximately 95% of the country's electricity is generated by gas.

Algerian Pipelines and Energy Lines of Communication

Algeria currently has four LNG plants, with a design capacity of 2.95 Bcf/d. Sonatrach's total upgrading program is anticipated to boost the country's LNG capacity to 3.29 Bcf/d. Algeria has special strategic importance, however, because its pipelines allow it to ship gas directly to Europe. These gas pipelines could also suddenly take on far more importance if any disruption took place in Russia's gas exports.

The 667-mile TransMediterranean (Transmed) pipeline, Algeria's first gas line, links the Hassi R'Mel gas field to Mazzara del Vallo in Sicily. The Transmed line *contains* segments *that pass* through Algeria (342 miles), Tunisia (230 miles), the Mediterranean (96 miles underwater) to Sicily and then on to Slovenia. New compression stations will enable the pipeline to carry up to 1 Tcf/y, its final capacity, to Italy and Slovenia. Most of the gas from this line is taken by Italy's main gas utility Snam, which is under contract to buy 680 Bcf/y from 1997 until 2018. Slovenia's Sozd Petrol is committed to 21 Bcf/y until 2007 under a contract signed in January 1990.

The $2.3-billion Gazoduc Maghreb-Europe (GME) pipeline began its first gas deliveries in November 1996. The pipeline is made up of five sections: 324 miles from Hassi R'Mel to the Moroccan border, 326 miles from Morocco to the Strait of Gibraltar, 28 miles across the Strait of Gibraltar at a depth of 1,312 feet, 168 miles from the Spanish coast to Cordoba, Spain where it ties into the Spanish transmission network, and 168 miles to Portugal.

It is unclear how much of a threat exists to either Algeria's oil or gas exports. Virtually all factions, including most Islamic extremists recognize that Algeria's economic growth and the welfare of its people depend on its ability to maximize exports. At the same time, blocking exports does gravely weaken any government and could serve the interests of an opposition. Furthermore, Algeria's ability to sustain and increase production is dependent on a stable mix of Algerian technocrats and foreign investment. A series of political and economic crisis could lead to sustained cuts in Algeria's output, rather than increases.

Libya

Libya is a classic example of both a "destabilizing " and a "failed state." It has been governed for nearly three decades by Colonel Muammar Abu Minyar al-Qadhafi, who overthrew the former monarchy in a bloodless coup in 1969. Qadhafi has borrowed ideas from various Islamic and pan-Arab movements to reject democracy and political parties and has tried to implement a "third way" superior to capitalism and communism. These ideas are outlined in Qadhafi's "Green Book." In theory Libya is ruled by the citizenry through a series of popular congresses, as laid out in the Constitutional Proclamation of 1969 and the Declaration on the Establishment of the Authority of the People of 1977. In practice Qadhafi and his inner circle control political power. Qadhafi uses extragovernmental organizations, including a Revolutionary Committee and a Comrades Organization, to exercise control over most aspects of citizens' lives.

Qadhafi is one of the most erratic military dictators in the world. He has challenged the West, Israel, and his neighbors at unpredictable intervals. He has tried to build up a major military machine, and Libya's arms imports have vastly exceeded its defensive needs and the forces it can properly man and support. It has contributed to terrorism throughout the world, provided endless regional clashes and tensions, invaded Chad, and has recently begun shipping tanks and other armor to Lebanon. He also uses extrajudicial killing and intimidation to control the opposition abroad and summary judicial proceedings to suppress opposition at home. The Government continues to repress banned Islamic groups and exercises tight control over ethnic and tribal minorities, such as Berbers and the Warfalla tribe.

Libyan Approach to Diversification: Energy Exports and Extremism

Libya's strategic importance lies largely in the fact that it is a significant energy exporter. Libya currently has 12 oil fields with reserves of one billion barrels or more, and two others with reserves of 500 million to one billion barrels. The U.S. Department of Energy estimates that Libyan production capacity will gradually increase from 1.5 million barrels per day in 1990-2000 to 1.7 million barrels per day in 2010, and then *fall back down* to 1.5 million barrels per day in 2020.

At the same time, Libya is a proliferator, a sponsor of terrorism, and a violent opponent of the Arab-Israeli peace process. Libya has been involved in five major clashes with U.S. forces — largely over Libyan claims to the Gulf of Sidra, whose waters Libya claims up to 32o 30' north, rather than abiding by the 12 nautical mile limit normally recognized in international law. The first clash occurred on March 21, 1973, when two Libyan Mirage fighters fired on a U.S.AF C- 130 reconnaissance aircraft patrolling a mission about 105 miles north of the Libyan coast. The C-130 turned away without damage. The second clash occurred on August 19, 1981, when one of two Libyan Su-22s fired an AA-2 missile at carrier based U.S. Navy F-14s that were demonstrating U.S. claims to freedom of navigation in the Gulf of Sidra. Both Su-22s were shot down. The third clash occurred on March 22, 1986, when an SA-5 unit located at Surt fired three SA-5 missiles at U.S. aircraft involved in similar freedom of the seas exercises. U.S. aircraft later destroyed the SA-5 radars using HARM missiles, as well as a Combattante-class patrol boat and Nanuchka-class guided missile patrol boat.

The fourth, and most serious, incident occurred on April 15, 1986, in retaliation for an attack by Libyan terrorists on the La Belle discotheque in the FRG on April 5, 1986. The attack killed three persons, including two U.S. servicemen, and wounded more than 200, many of them seriously. In opening remarks, the German prosecutor said the bombing was "definitely an act of assassination commissioned by the Libyan state." German authorities issued warrants for four other Libyan officials for their role in the case who are believed to be in Libya.

U.S. F-111's based in the United Kingdom, and U.S. carrier based aircraft bombed the command center at the Al Azizyah Barracks and airfields in Benina and Tripoli. A number of Libyan IL-76 and MiG-23 aircraft were destroyed, and one F-111 crashed at sea. The most recent incident occurred on January 4, 1989, when two U.S. Navy F-14s operating off the northeastern coast of Libya detected Libyan MiG-23s tracking them with their radars. Both MiG-23s were shot down by the F-14s using Sparrow and Sidewinder missiles.

These U.S. raids and victories in air-to-air combat led Qadhafi to adopt a policy of partial conciliation with the West, but they also led to new acts of terrorism. On December 21, 1988, Libya planted explosives on an Air Malta flight connecting with Pan Am Flight 103 from Frankfurt, Germany to the U.S. The plane blew up over Lockerbie, Scotland, killing 280 persons. Libya also planted a bomb on a French UTA airliner, Flight 772, in 1989 — perhaps in retaliation for France's support of Chad. The airliner blew up over the Sahara and killed another 170 people.

UN and U.S. Sanctions

On April 15, 1992, the United Nations imposed economic sanctions on Libya for refusing to extradite two Libyan nationals accused of carrying out the *bombing of Pan Am Flight 103*. These sanctions included the grounding of all air traffic to and from Libya, a reduction in diplomatic relations and a ban on all arms sales to the country. As of early November 1997, Libya had not surrendered the two bombing suspects, despite indications that sanctions were adversely affecting Libya's economy. In late September 1997, the Arab League and the Non-Aligned nations both called for an easing of sanctions on Libya. The United States responded that *sanctions would not cease* unless Libya turned over the bombing suspects.

UN sanctions were expanded in November 1993 to include a freeze on Libyan funds overseas, a ban on the sale of oil equipment for oil and gas export terminals and refineries, tougher restrictions on civil aviation, and the supply of arms. The United States, which enforces its own sanctions against Libya, had made further efforts to convince the UN to consider a total embargo on Libyan oil. However, several key European allies that rely heavily on Libya's low sulfur oil, particularly Italy (which gets 80% of its oil from Libya) and Germany, expressed deep reservations about this option.

The United States imposed additional sanctions on Libya on August 5, 1996. This action—the U.S. Sanctions Act of 1996 — extends U.S. sanctions on Libya to cover foreign companies that make new investments of $40 million or more over a 12-month period in Libya's oil or gas sectors. UN sanctions forced Libya to adopt a more conservative fiscal policy and to limit public infrastructure spending to a few main projects, such as the Great Man Made River (GMR), a $25 billion project to bring water from underground aquifers beneath the Sahara to the Mediterranean coast. The agricultural sector is a top priority for the Libyan government, which hopes that the GMR will reduce the country's water shortage and its dependence on food imports.

This situation has been partially resolved. After months of negotiations and brokering by Saudi Arabian Ambassador Prince Bandar bin Sultan, South African President Nelson Mandela, Egyptian President Hosni Mubarak, and UN Secretary-General Kofi Annan, President Qadhafi released the two suspects on April 5, 1999. They are standing trial at a former air base in the Netherlands under Scottish law.

In return for releasing the suspects, UN sanctions on Libya were suspended for ninety days, after which Secretary-General Annan reported to the Security Council on Libya's cooperation. If Libya complied with other conditions established by the Security Council, sanctions could be lifted altogether. Although it agreed to the suspension of UN sanctions, the United States has declared that it will not lift U.S. sanctions or remove Libya from its official list of terrorist states until Libya complied with all the provisions of UN resolutions on Libya. The United States requested face-to-face talks with Libya so that they could decide whether to support the permanent lifting of sanctions.

Secretary General Annan reported his findings on Libya to the Security Council on July 2, 1999. According to Annan's report, Libya has, so far, complied with the provisions set forth by the Security Council. France has been satisfied with Libya's cooperation in regards to UTA Flight 772, and has convicted 6 people in absentia for the crime. Libya has provided assurances that it will continue to comply with the remaining requirements, including the compensation of victims' families in the event of a guilty verdict in the Pan Am 103 trial. The United Kingdom has stated that it has been satisfied with Libya's cooperation in combating terrorism, and hopes that Libya will continue to renounce terrorism. The United States, on the other hand, has not been as pleased with Libya's progress, and it does not seem likely that UN sanctions will be completely lifted any time soon.

The long years of sanctions have helped fuel Islamic unrest and sectarian divisions, and created growing social and economic problems. Partly due to sanctions, Libya's government is moving ahead on a plan to share oil wealth with the country's citizens, particularly those with low incomes.

Sanctions and Energy

The Department of Energy estimates that Libya's ability to increase its oil production (and exports) apparently has been hampered by sanctions - although to what degree is uncertain—mainly due to a ban on needed enhanced oil recovery equipment. Despite sanctions, Libya exported about 1.1 million bbl/d of crude oil and 200,000 bbl/d of products before the current crisis in oil prices.. Nearly all (about 90%) Libya's oil exports are sold to European countries like Italy (580,000 bbl/d in 1996), Germany (258,000 bbl/d), Spain and Greece.

The suspension of sanctions may offer Libya the means to revitalize its older fields. Sanctions have prevented the development of older fields run by NOC by banning the import of spare parts needed to maintain the fields and enhance production. Several fields operated by NOC affiliates are suffering from low pressure, which has held back production. The procurement of equipment for these fields is likely to be a priority. However, the situation is complicated by the fact that most of these fields were originally developed by U.S. firms and will need U.S.-manufactured spare parts.

The decline in Libya's older fields has been offset by fresh discoveries by European operators. Spain's Repsol is now producing about 140,000 b/d in the Murzuk basin in the southwest, and announced discoveries in early 1999 that indicated potential reserves of between 100-200 million barrels. British firm Lasmo is developing its Elephant discovery in the same region and is due to produce the first oil there in 2000, with output rising to 150,000 b/d in 2002. Other international groups active in the upstream oil sector include Italy's Agip, Canadian Occidental, Canada's Red Sea Oil Corporation, Dublin-based Bula Resources, and France's Elf Aquitaine.

Oil export revenues account for about 95% of Libya's hard currency earnings. In 1997, oil production stood at 1.43 million bbl/d, down from over 3 million bbl/d in 1970. Libya would like to boost production, but sanctions *have caused* delays in a number of field development and enhanced oil recovery projects, as well as *deterred* foreign capital investment. Faced with a mature oil reserve base, Libya's challenge is *to maintain* production at older fields while at the same time bringing new fields online. Reserve replacement, however, has been slipping since the 1970s.

The geopolitics of sanctions also *affected* Libyan gas production. The Department of Energy estimated Libya's proven natural gas reserves at 46.3 trillion cubic feet (Tcf) in 1997. Libya believes the country's actual gas reserves to be considerably larger, possibly 50-70 Tcf. Large new discoveries have been made in the Ghadames and el-Bouri fields, as well as in the Sirte basin.

Continued expansion of gas production remains a high priority for Libya for two main reasons. First, Libya has aimed to use gas instead of oil domestically, freeing up more oil for export. Second, Libya is looking to increase its gas exports. Libya also produces a small amount of liquefied petroleum gas (LPG), most of which is consumed by domestic refineries. Natural gas consumption has been rising at a 10% annual rate since 1990. Besides gas used for injection into oil fields, most of this consumption has been by the petrochemical industry at Ras Lanuf, and by electric power sectors. In recent years, several power plants have switched from fuel oil to natural gas, and four new gas-powered plants have been built recently.

The Department of Energy reports that considerable potential exists for a large increase in Libyan gas exports to Europe. In June 1996, Italy's ENI's group signed an agreement with Libya's state-owned National Oil Company on a plan whereby Italy will import around 280 Bcf of Libyan gas per year by pipeline under the Mediterranean to southeastern Sicily and then on to the Italian mainland. Total investment in this project is expected to be around $5.7 billion, and gas is expected to begin flowing in 2000.

Libyan Pipelines and Energy Lines of Communication

Agip also has promoted linking the reserves of both Egypt and Libya to Italy by pipeline. An agreement in principle to link Egypt and Libya's gas grids was reached in June 1997, following a visit to Libya by Egyptian President Husni Mubarak. Yet another proposal is to build a pipeline from Egypt and Libya to Tunisia and Algeria, from where it would hook up with the existing pipeline to Morocco and Spain.

Tunisia and Libya agreed in May 1997 to set up a joint venture which will build a natural gas pipeline from the Mellita area in Libya to the southern Tunisian city and industrial zone of Gabes. In 1971, Libya became the second country in the world (after Algeria in 1964) to export liquefied natural gas (LNG). Since then, Libya's LNG exports have generally languished, largely due to technical limitations which do not allow Libya to extract LPG from the LNG, thereby forcing the buyer to do so

The Arab-Israeli Peace Process

There are other energy exporting states in the Middle East, *including* Egypt, Syria, and Yemen. Their reserves and exports are so limited, however, that they have negligible geopolitical impact. The two issues that are likely to have major geopolitical impact are the future of the Arab-Israeli ring states and the role of U.S. policies and power projection capabilities in stabilizing the region. There also, however, are potential threats to key chokepoints affecting oil exports out of the MENA region.

The Arab-Israeli ring states – Egypt, Israel, Jordan, Lebanon, and Syria – are a large geopolitical problem between two large pools of oil reserves. No geopolitical analysis of the region, however, can ignore the impact that the Arab-Israeli conflict has had in terms of creating tensions between the West and the Arab and Islamic worlds when there are delays or problems in the peace process. No analysis can ignore the fact that the U.S. has a firm strategic commitment to Israel, that there are no condition under which it will not act to prevent an existential threat to Israel, and that the U.S. must be prepared to live with any negative consequences in terms of Arab and Islamic reactions.

Egypt remains the most influential Arab nation, and the key to maintaining an Arab-Israel peace. It is the only Arab military power large enough to intervene effectively in the region, and controls the Suez Canal and the SuMed pipeline. Egypt's secular regime faces serious economic challenges, although it is moving towards successful economic reform. It also faces a challenge from Islamic extremists. It seems unlikely that Egypt currently faces a crisis in terms of internal instability, but a major breakdown in the peace process and/or global recession might create far more serious problems than exist today.

Jordan is another key to a successful Arab-Israeli peace, and plays an important role in securing Western Saudi Arabia and in containing Iraq. King Hussein's sudden shake-up of the line of succession just prior to his death in February 1999, led to initial concern over the future stability of the nation.

His brother Hassan had been Crown Prince for 34 years, and was well known in the Middle East. However, King Hussein suddenly replaced Prince Hassan with his eldest son Abdullah who was far less well known. This prompted considerable concern over Abdullah's ability to effectively govern the nation. Though it is too early to tell, it now appears as though Abdullah has a firm base of power and that the transition of authority has gone smoothly. Still, Jordan faces severe economic problems and uncertainty about the Arab-Israeli peace process. In addition, there are strong Islamic and Palestinian factions that do not support its monarchy and are pushing for radical reforms. Therefore, it is possible that Jordan could enter into a period of prolonged instability.

Syria's differences with Israel over the Golan and Lebanon are a major factor blocking the completion of the peace process, and pose the most serious threat that a new Arab-Israeli conflict may occur. Syria is acquiring long range missiles, VX gas, and developing biological weapons. Syria's state-dominated economy and high birth rate are seriously threatening its development. At some point between 1996 and 2010, Syria will face a succession crisis when it must replace Hafez Assad. This could trigger serious tension or strife between its Alawite minority, secular Sunnis, and the remnants of the Muslim Brotherhood.

Lebanon is the scene of a proxy war between Israel and Hezbollah forces backed by Iran and Syria, and its ambitious recovery and development plans seem to be failing to produce the kind of results that can bring stability. Serious questions exists as to whether Syria will give up its control of much of Lebanon, and as to whether Lebanon can avoid another round of sectarian and ethnic conflict.

Much of the stability of the peace process and the Arab-Israeli ring states depends on the emergence of the Palestinian Authority as a sovereign entity and the transformation of that Authority into a successful political and economic structure that can meet the needs of its people, minimize Islamic extremism, and put an end to terrorism. The Palestinian Authority is now caught up in the crisis over the peace process. It faces a major succession crisis in replacing Yasser Arafat, and it has had only limited success in evolving a capability to govern. Its economy is paralyzed by tensions with Israel and internal mismanagement and corruption. This presents serious risks for the peace process, in terms of terrorism, and the future stability of a Jordan with a large Palestinian population.

Impact of U.S. Policy and Power Projection Capabilities

There are many nations outside the Middle East that heavily impact its future. *These* include immediate neighbors like Turkey, Afghanistan, Pakistan, Eritrea and Ethiopia. *They also include* peripheral powers like Russia, India, the Caspian and Central Asian states, and key importers like the EU and Asia. Asia, in particular, is becoming a steadily more important consumer of Middle Eastern oil and will probably consume about 70% of all Gulf exports by the year 2010.

Security, however, is heavily dependent on the ability of U.S. power projection forces to enforce deterrence and stability in the Gulf, deal with the threat of proliferation, and enforce conflict termination in the face of outside aggression. It is also dependent on the success of U.S. policy in dealing with Iran, Iraq, and the Arab-Israeli peace process.

The basic U.S. power projection strategy in the Gulf and MENA region seems sound. It is, however, badly underfunded, and forces are over-stretched and over-deployed. The U.S. also has problems in dealing with Iraq. There is nothing wrong with pursuing sanctions and inspections as long as it is balanced by a strong oil for food program. There is nothing wrong in supporting the Iraqi opposition – although this seems unlikely to have much impact. The U.S. *does* not, however, *have a* credible end game for dealing with Saddam Hussein.

The U.S. should do everything it can to limit the pace of proleferation in Iran, and to halt the transfer of advanced military technology. However, the U.S. is moving too slowly in its efforts to reach out to Iran's moderates and the kind of economic links and energy investment that could have a positive impact. The backlash from the Arab-Israeli peace process is combining with Iraq's success in exploiting and exaggerating the "hardship" issue and creating serious anti-American sentiment even in friendly Arab states. Many experts feel that the U.S. is not conducting an effective campaign to explain its positions and win friends in the region.

Red Sea: A Geopolitical Threat

It is too soon to describe the Red Sea as a serious geopolitical threat to the flow of oil. Nevertheless, there is a risk that the instability in states like Egypt, Eritrea, Ethiopia, Somalia, the Sudan, and Yemen could affect tanker traffic through the Red Sea, and Saudi or Yemeni shipment out of ports in the Red Sea. Libya covertly mined the Red Sea in 1982. The Sudan is a radical state that poses a terrorist threat to its neighbors. Yemen is involved in low level fighting over islands in the Red Sea with Eritrea and Saudi Arabia. Eritrea is fighting a border war with Ethiopia, Djibouti and Somalia are the scene of tribal conflicts, and Egypt still faces a low level problem with Islamic Extremists.

The Suez Canal

Egypt connects the Red Sea and Gulf of Suez with the Mediterranean Sea. This currently involves the flow of 3.5 million b/d of oil (1.1 million b/d through Suez Canal, 2.4 million b/d through Sumed Pipeline). The principal destinations are Europe and the United States. Closure of the Suez Canal and/or Sumed Pipeline would force tankers to go around the southern tip of Africa (the Cape of Good Hope), and increase transit time and cost.

The Bab al-Mandab

Similarly, it is possible that shipping through the Bab al-Mandab (the narrow gateway to the Red Sea) could be disrupted. This could *prevent Gulf oil* from reaching the Suez/Sumed complex. Yemen fought a brief battle with Eritrea over Greater Hanish island, located just north of the Bab al-Mandab in December, 1995. Saudi Arabia and Yemen clashed over another island in 1998.

Two Red Sea states require special attention. Yemen is attempting to create a more democratic regime, but it faces serious challenges because of chronic mismanagement of its economy, massive population growth, divisions between the north and south, and tribal and sectarian divisions. Yemen has serious border disputes with Saudi Arabia, and Saudi Arabia is deliberately encouraging tension within Yemen to weaken its government. This affects the development of oil resources in the border area and the rest of Yemen, and poses the risk of more serious future conflicts between Yemen and Saudi Arabia. These are unlikely to have any direct impact on Saudi energy production and exports, but could add to Saudi Arabia's internal problems.

The Sudan has significant oil resources, but years of civil war have blocked their development and efforts to establish a reliable estimate of proven reserves. The Sudan has an Islamic extremist regime that threatens its neighbors and has supported terrorism.

References

CIA, World Factbook, 1997, Washington, GPO, CD ROM, "Saudi Arabia;" The World Bank, World Bank Atlas,1998, Washington, World Bank, pp. 36-37.

Energy Information Agency, 1999, *International Energy Outlook*, Washington, DOE/ EIA-0484(99), March 1999, p. 32.

Petroleum Finance Corporation, 1998, "Saudi Oil Policy Options – Time to Choose: Market Share or Price Defense?,"Washington, December.

U.S. Arms Control and Disarmament Agency (ACDA), 1996, *World Military Expenditures and Arms Transfer*, GPO, Washington.

World Bank, 1998, World Development Indicators, Washington, World Bank, p. 14.

8

Energy Politics and the Future

The Middle East does not have a single future. There are twenty-three countries in the region and few similarities between the major oil powers discussed earlier. It seems almost certain that some countries will succeed, most will muddle through, and some will either fail to govern or fail to properly develop. There are, however, significant pressures on several key energy exporters. As has been discussed earlier, *four* critical countries – Algeria, Iran, Iraq, and Libya – have very uncertain futures. Saudi Arabia, the most pivotal oil power in the world, is under considerable economic stress. At the same time, about the only thing the oil rich countries in the region have in common at this point is declining per capita income, a lack of economic diversification, serious demographic pressures, and political uncertainty.

There is always a great deal of literature that provide estimates of how the Middle East might develop, and both individual countries and many international institutions provide estimates that seem to show significant growth in recent years. Four decades of development literature and reform plans, however, have produced surprisingly little change, and it is all too clear that most estimates of current and near growth are exaggerated.

A World Bank analysis of world economic trends over the 30-year period from 1965 to 1997 *showed that* the Middle East and North Africa averaged 3.1% real annual growth in GNP, versus 3.1% for high income states, 4% for middle income states, and 3.8% for low income states. The World Bank estimates that average per capita income grew by only 0.1%

from 1965 to 1996 — a period of more than three decades — while the total population increased by an annual average of 2.8% and the labor force increased by 2.8%. .

Conventional forecasting models have been about as accurate for the MENA region as stockbroker picks for the stock market. It does seem clear, however, that currently projected declines in the birth rate will come too late to substitute for economic reform. The MENA region had a population of 175 million in 1980. It was 376 million in 1996.

Even if average annual population growth drops from 2.9% during 1980-1996, to 2.1% during 1996-2010, the population of the MENA area will still grow to 376 million. There is also nothing typical about this population growth. Low income nations as a whole exhibited 2.3% population growth during 1980-1997, and are projected to grow by 1.7% during 1997-2010. The averages for low and middle-income states are 1.8% and 1.3% respectively, and the averages for high-income states are 0.7% and 0.3%.

Moving Towards Growth and Stability

That said, the Middle East scarcely faces a Malthusian nightmare. Moderate political leadership, basic economic reforms, and aggressive efforts to reduce population growth and/or over-dependence on foreign labor could move virtually every Middle Eastern state towards sustained real economic growth in terms of both its GDP and per capita income within three to five years. Rough working estimates by the World Bank indicate that sustained growth could average twice the population growth rate in half a decade *–although*, these estimates were made at a time when it was assumed that Asian economic demand would keep oil prices moderate to high.

Reform would be easiest to implement in the Southern Gulf, and would affect three key oil-exporting states: Kuwait, Saudi Arabia, and the UAE. Iran would require a definitive shift towards political moderation to implement such policies, and Iraq would require not only the replacement of Saddam Hussein, but a fundamentally different form of leadership elite that at least *would* provide equity to Iraq's Shi'ite majority and Kurdish minority.

North Africa presents more challenges. Algeria would have to solve its political crisis and put a firm end to its civil war. Libya would require both a fundamentally different leader and a leadership grounded in reality.

North Africa has much less average oil and wealth per capita, and its average per capita income has dropped from a peak of $2000 in 1985, in constant 1995 dollars, to around $1,600 in 1997. By contrast, the average per capita income has dropped from a peak of $3,600 in 1985 to around $2,700 in 1997. Morocco and Tunisia have, however, at least begun the process of political and economic reform, and Algeria's plans are a beginning. The Egyptian private sector is growing steadily, and is an indication of just how much the right kind of reform might accomplish.

The Arab-Israeli states are peripheral to the geopolitics of energy, but much still depends on the success of the peace process. There is enough political linkage between the peace process, and the ability to concentrate on development in the rest of the region, that a best case *scenario* would require steady progress. At the same time, there must be coherent economic reform. Israel may be much wealthier than its Arab neighbors, a strong and vibrant private sector, but state and political interference and a swollen public sector are as much of an enemy to Israel as Iran or any Arab state. Jordan has talked and planned reform, but has not implemented it. Syria *is mired* in a state-dominated past. The Palestinians lack governance and have no real economy. Lebanon is a glorified Ponzi scheme – all construction and infrastructure, and no real productivity and exports. Major changes would be required in the leadership and political elites of at least Syria and the Palestinian Authority.

It should be noted that a reform and development scenario has several other requirements. *Natiions must avoid new* major conflicts or civil wars, and military expenditures and arms imports *must* remain relatively low. (They have shrunk from about 17% of GNP in 1985 to around 8% today, and from around $28 billion in 1985, in constant 1995 dollars, to around $14 billion). The other states of the Middle East must follow Iran's lead in making aggressive efforts to reduce population growth – which ultimately may have to be kept well below 2%. The region must privatize and reform in ways which allow countries with minimal foreign investment to attract well over $1 trillion in added private and foreign investment between 1998 and 2010, at least $500 billion of which must go into energy investment, infrastructure, utilities, and water.

At the same time, the West needs to be very careful about vacuous ideological nonsense that states peace, democracy, human rights, and/or reductions in corruption will solve the region's basic problems. These are important human values, and progress *in these areas* may well be necessary to achieve progress in other areas. The previous analysis has shown, however, that *the* key structural changes necessary to achieve the "best case" are matters of economics and demographics. In fact, there is little about the history of the MENA region to indicate that it matters very much whether a country has a president, dictator, socialist, king, or sheik. The results are remarkably similar in terms of overall development and per capita wealth.

Muddling Through

The problem with the muddling through scenario, which is *likely to be* the real-world case for most states, is that it is national, not regional. Some countries – Israel, Kuwait, Qatar, and the UAE – will probably either make enough progress to partially resolve their development problems or earn enough oil wealth to minimize their impact. Egypt, Iran, Morocco, Oman, Saudi Arabia, and Tunisia require substantially better leadership and efforts at reform, but can probably muddle through with some degree of growth or minimal further strain. Algeria, Bahrain, Lebanon, and Syria are more marginal cases, politically, economically, and demographically. They are likely to muddle downwards as muddle through. Iraq, Libya, the Sudan, and Yemen are high-risk cases.

Muddling through does not imply that there will not be low intensity and civil conflict, terrorism, and individual failed states. It does mean, however, that regional states will continue to be desperate to maximize oil and gas export revenues, and will invest heavily in expanded production. There may be occasional blips or interruptions in some aspect of oil exports, and it is questionable whether sufficient capital will go into development to reach the DOE estimates of oil production listed earlier. In broad geopolitical terms, however, the MENA region would be a reasonably reliable supplier over time, and become even more dependent on imports from outside the region - - ensuring a need to recycle petro-dollars or petro-yen.

It is also unlikely that the nations that follow this scenario will force a rate of increase in oil and gas prices in constant dollars that *will be* serious enough over time to reduce the growth in global GNP. There may well be the kind of limited and crisis-driven cuts in supply which would present sporadic problems for the world economy. "Muddling through," however, means any conflicts will be limited enough to have minimal impact on the key exporting states. The question for energy security is whether "muddling through" will provide enough new investment and energy capacity to maintain the world's swing production capacity, and keep demand from outstripping surplus capacity over time. Further, if the Middle East is dominated by states that " muddle through," the resulting increases in energy production capacity will probably make the global economy more vulnerable if supply problems occur in any other key regions – such as the FSU – or the demand for oil increases because of supply problems in other energy sources like Chinese coal or European and Japanese nuclear energy.

FAILED STATES

True worst case scenarios are pointless. Not only is everyone already dead in such scenarios, they have never been born. In the case of the MENA region, these worst case scenarios include the prolonged internal collapse of a major exporter like Saudi Arabia, wars that cause prolonged total interruptions in all Gulf oil exports, nuclear or WMD attacks on critical exporting facilities, plus the odd meteor or earthquake. Such futures are possible – particularly over a period as long as 1998 to 2020. However, such futures are scarcely *probable* and they virtually defy cost-effective contingency planning.

Other risks are more probable and they are still quite serious. They can cause the kind of energy "crisis" which may never make a movie or novel, but which can have a slow and cumulative effect in raising real energy prices and slowing global economic growth. They can cause serious human tragedies in parts of the MENA region, and lead to serious short term panics or cuts in growth.

These cases include a mix of events in which regional states fail to deal with their internal problems, become involved in serious regional conflicts, or "institutionalize" problems that block effective investment and development of their resources. These cases include:

- A systematic, long-term break down in the Arab-Israel peace process with a backlash that limits U.S. ability to deter and terminate conflict in the Gulf, increases terrorism and Islamic extremism throughout the region, and limits outside investment.
- The failure of Iranian political and economic reforms, and of Iran's oil and gas development.

- Another round of Iraqi aggression, threatening neighbors like Kuwait and preventing or reducing Iraq exports and increases in export capacity.
- Internal civil conflict or tension in Saudi Arabia, or a shift to a kind of Islamic extremism that discourages further increases in oil export capacity and tries to use oil exports as a political weapon.
- State sponsored or proxy terrorism, possibly using weapons of mass destruction, of a kind that would have a major impact on energy investment and/or the willingness and capability to export.
- The kind of regional proliferation that would force a major arms race and create a structure of deterrence sufficiently uncertain to make investment a major risk, possibly coupled to some actual use of a few weapons against an oil-exporting country(ies), with the resulting panic.
- Prolonged civil war in Algeria and/or Libya that affected exports or blocked effective investment and development of export capacity.
- Naval/air conflicts in areas like the Strait of Hormuz that had a serious short-term impact on the flow of tankers.
- Prolonged U.S. economic sanctions which affect energy investment in key exporting states.
- A failure by the West to sustain the necessary power projection forces, regional political pressures that make them difficult or impossible to use, and a regional failure to develop any effective regional security structure.

It is important to realize that the risks in such a "worst case" scenario are not that any one problem is probable in the form listed above, but rather the cumulative probability that one or more event of the *kind* listed above will occur over a period as long as 1998-2010. This kind of uncertainty is not a reality that planners and forecasters like, but it is a mathematical fact of life – and a constant historical reality. When it comes to risk assessment in the Middle East, the cumulative probability that one of many less probable events will actually occur over time is generally higher than the probability that one of a small set of more probable events will occur.

Geopolitical Time Bombs

There are no "ticking time bombs" in the sense that any given threat is already certain to explode. However, there are cases that deserve special attention.

Saudi Arabian Economic Reform and Succession Issues

Saudi Arabia can almost certainly make an easy transition from Kind Fahd to King Abdullah. In fact, Prince Abdullah is probably better equipped to control the budget, push economic reform, and prevent Islamic fundamentalism from becoming Islamic extremism. Saudi Arabia does, however, face serious structural challenges in its economy and demographics and future successions may be more difficult.

Saudi Arabia's per capita income has already dropped to around $6,500 in constant $U.S. 1995 dollars, versus well over $10,000 at the peak of the oil boom. It is encountering serious, prolonged budget deficit problems, and oil revenues are only about 60% of their total in 1997. Saudi Arabia experienced a 3% drop in real per capita income during 1995-1996 – when oil revenues were relatively high.

Like Oman and many other Arab states, Saudi Arabia finds it *very difficult* to come to grips with population growth, which the World Bank estimates at an average of 4.4% during 1965-1996, and as rising to 4.6% during 1980 to 1996. Even if this rate drops to 3.3% during 1996-2010, Saudi Arabia's population will have risen from 9.4 million in 1980 to 20.1 million in 1997, and will rise further to 35 million by 2015. This *is* a ticking demographic time bomb.

It is far too soon to predict any crisis, but Saudi Arabia is so pivotal in geopolitical energy terms that even a possibility remains a serious risk.

Iran's Political Stability and Energy Development

Iran may be moving towards moderation, but military power remains in the hands of the traditionalists and hard-liners, Iran continues to proliferate, build-up its forces in the lower Gulf and Gulf of Oman, to support the Hezbollah, and preserve its infrastructure for terrorism. The Khatami regime faces further challenges in terms of Iran's economy, and these are compounded by U.S. economic sanctions – which seem far more likely to encourage extremism and hostility than aid Iran's moderates.

Post-Saddam Iraq

The U.S. has no probable "end game" in dealing with Iraq, and its new posture of "overt opposition" to the Iraqi regime is hollow rhetoric. Oil for food can buy time in enforcing sanctions that stop most Iraqi arms imports and imports of dual-use technology, but UNSCOM and IAEA inspections are a rapidly wasting asset. Like it or not, the U.S. will probably have to live with a revanchist Saddam who will slowly break out of sanctions and acquire significant numbers of biological and chemical weapons.

Kuwaiti Security

Kuwait is only marginally more effective as a military power than it was in 1990. Saudi Arabia's military modernization and expansion is successful only at the tactical level, and Saudi Arabia has almost totally failed to create the effective land forces needed to aid Kuwait or provide forward defense of its foreign border. The rest of the GCC has token reinforcement capabilities at best, *thereby making* Kuwait's vulnerability one of the defining geopolitical risks of the Middle East.

The Kurdish Problem

The Kurd areas between Turkey and Iraq are already the scene of a Turkish civil war involving a reinforced corps-level force of Turkish troops and clashes that constantly spill over into Iraq. The Kurdish enclave in Iraq is the scene of a low intensity conflict between two warring Kurdish factions (Barzani versus Talibani), and Saddam Hussein sees the break up of the Kurdish security zone as a major objective. There is a serious risk this could trigger a new civil conflict in Iraq, with some impact on energy development in Northern Iraq. It could also trigger a new major crisis between Iraq and the United States.

Algerian Political Stability

The bloody civil war in Algeria *continues* with only limited reform and uncertain efforts to reach some compromise or stable form of conflict resolution. There is a constant possibility that the targets of the Islamic extremists may be expanded to include energy facilities, mass attacks on Westerners, or attacks designed to prevent investment that strengthens the government. The military junta that is the true power behind the presidential façade is almost as *dangerous* as the extremists, continues to be corrupt and divisive, and probably cannot implement Algeria's economic reform plans.

Qadhafi's Legacy

Qadhafi is creating a legacy of Islamic fundamentalism, Arab-Berber tension, and slowed development which can seriously erode Libya's near-term energy development and *pave the way for* serious internal civil conflict *and instability.*

The Arab-Israeli Peace Process

The linkage between the Arab-Israeli peace process and regional support for the U.S. is currently eroding support for the U.S. in the Southern Gulf, Arab-Israeli ring states, and North Africa. The backlash is particularly serious among intellectuals and the media in the Gulf, where it combines with the feeling *that* the U.S. has imposed unnecessary suffering on the Iraqi people, has forced the GCC states to buy arms they do not need, and is acting as a regional mercenary.

Proliferation

Proliferation is a growing problem in the Middle East, and one that is not likely to diminish in the near future. There is a complex pattern of proliferation in the region, and the range of delivery systems is steadily expanding. Algeria, Egypt, Iran, Iraq, Israel, Libya and the Sudan are all involved in acquiring weapons of mass destruction. Israel has large nuclear forces, Iran has a major nuclear program, Algeria seems to have a contingency program, and Iraq preserves the technical skills necessary to make nuclear weapons. Iran, Iraq, and Syria have significant biological warfare programs. Egypt, Iran, Iraq, Israel and Saudi Arabia have long-range missiles or programs to acquire them.

Guessing at Risk and Priorities

These are so different in terms of regime, goals, and behavior that it is obvious that there is no regional threat to the West, but rather the possibility that individual states might pose a threat to individual Western nations or interests. Three major proliferators – Iran, Iraq, and Libya – are of special interest. These are nations that have posed a threat to the West in the past and which have also sponsored attacks of state terrorism against Western targets/and or on Western soil.

Iran currently poses the most significant near-term *threat* in terms of acquiring biological and nuclear weapons, and long-range missiles that might strike Europe or the United States. In spite of Iranian denials, there is little doubt that Iran has an active nuclear and biological weapons program, and has already begun to test long-range missiles. Iran's capabilities, however, will remain highly limited for the next decade, and Iran faces a strong regional threat from Israel. While Iran's regime may or may not become truly moderate in character, it has become progressively more pragmatic since the death of Khomeini, and it is far from clear that it would take "existential" risks of the kind posed by such an attack on the West.

Libya has the dubious distinction of being the only Middle Eastern state to have fired a long-range missile on a Western target – it fired on the Italian island of Lampadusa following the U.S. raid on Tripoli. At the same time, Libya's grandiose military plans have ended in failure. Libya has some chemical weapons capability, but has failed to develop ballistic missiles with longer ranges than the Scud. It has explored biological and nuclear weapons programs, but there is little evidence of success.

Iraq's has made massive efforts to proliferate and to acquire long-range missiles, and biological and nuclear weapons. These programs would already pose a serious potential threat to the West had *they* not been halted by the Gulf War, and by the efforts of UNSCOM and the IAEA. Most of Iraq's past capabilities have been largely destroyed.

No Middle Eastern state can disregard the fact that any use of a biological or nuclear weapon that produced massive casualties could trigger devastating conventional strategic strikes or even the use of nuclear weapons by the West.

At the same time, there are dangers in assuming that Middle Eastern states will always behave as "rational actors." The history of the region is filled with miscalculations, erratic behavior, and risk taking. Behavior can alter rapidly in a crisis, and the most threatening states have rulers or ruling elites that may choose to escalate in ways that are far less conservative than *what* Western planners would *consider* under similar conditions.

The following scenarios may not represent even moderate probability cases, but they are possible enough so they deserve serious consideration:

- Weapons of mass destruction might be used against key energy and energy export facilities in intra-regional conflicts, posing a major economic threat.
- Attacks might be carried out on Western power projection forces in the region, or the threat of such attacks might be used to try to force a regional power to expel Western power projection forces or carry out other acts hostile to Western interests.
- Threats against the West, demonstrative long-range missile attacks against targets in the West, or low-level use of weapons of mass destruction might be used to try to force Western nations to support

the policies of a given Middle Eastern state, or intervene in a regional conflict. The escalation of an Israeli-Syrian conflict, or future Iranian-Iraqi conflict might lead to such a threat.

- A regional power might set up a launch-under-attack system targeted on the West in an effort to deter Western intervention or military action. Such a system might be created to prevent Western counterproliferation strikes.
- The threat, demonstrative use, or larger scale use of such weapons *might be utilized in an* effort to force an end to sanctions.
- A regime on the edge of collapse might lash out, feeling it had nothing to lose and accepting the risk of broader retaliation against the nation. Alternatively, a nation under nuclear attack by Israel might feel that attacks were justified against Western targets, particularly U.S. bases.
- Middle Eastern states are not limited to conventional forms of warfare. While a great deal of attention focuses on long-range missiles, a Middle Eastern state might use unconventional delivery means or a terrorist proxy to deliver such weapons – hoping that it would not be identified as the source or that enough ambiguity would exist to prevent a decisive response.
- Technology or fissile material transfers might suddenly destabilize the balance. This might include the transfer of long-range missiles or fissile material, or key components and technology for missiles and weapons. This could suddenly alter the regional balance and the perceived risk in threatening the West or Western interests.

Once again, the problem is to balance possible risks against probable risks, and draw suitable consequences for policy. The actions of most Middle East states and leaders are normally cautious and self-preservation is normally the highest single priority. There is no question, however, that a combination of creeping proliferation, and relying on the judgment and stability of at least five major proliferators, presents risks.

Terrorism

Middle Eastern terrorism remains a very real threat to the moderate regimes in the Middle East and a transnational threat to the West. There have been very serious terrorist attacks on Western targets in the Middle East in the past, such as the bombing of the Marine Corps Barracks in Beirut. There *has been a history* of bloody attacks outside the Middle East, such as the bombings of Pan Am flight 103 over Scotland in 1988 and the bombing of UTA flight 772 over Chad in 1989. The bombing of Pan Am flight 103 killed 259 people on board and 11 people on the ground, and the bombing of UTA flight 772 killed 171 people on board.

Terrorism is also a threat which may be sharply escalating. Middle Eastern attacks on Western targets dropped in frequency and intensity (casualties) after the Gulf War and Israel's peace agreement with the Palestinians. This pattern may be reversing. The bombing of the National Guard Headquarters and Al Khobar towers in Saudi Arabia, however, seems to have been the harbinger of much larger terrorist *activities*. The November 13, 1995 truck bombing of the National Guard Headquarters in Riyadh killed five U.S. service men and two Iranians.

The June 25, 1996 bombing of the Khobar Towers killed 19 U.S. servicemen, and it involved a massive device. The attacks on the U.S. Embassies in Kenya and Tanzania involved massive numbers of innocent casualties – 247 dead and over 5,000 wounded in the case of Kenya, and 10 dead and more than 75 wounded in the case of Tanzania. These attacks involved truck bombs with 600-800 pounds of explosive.

Civil tension in the Middle East has made tourists a growing target. For example, the worst terrorist attack in Egypt's history occurred on November 17, 1997. Six gunmen belonging to the Egyptian terrorist group al-Gama'at al-Islamiyya (Islamic Group or IG) entered the Hatsheput Temple in Luxor. For nearly half an hour, they methodically shot and knifed tourists trapped inside the Temple's alcoves. Fifty-eight foreign tourists were murdered, along with three Egyptian police officers and one Egyptian tour guide. The gunmen then fled the scene, although Egyptian security forces pursued them and all six were killed. Terrorists launched a grenade attack on a tour bus parked in front of the Egyptian National Antiquities Museum in Cairo on September 18, 1997, *killing* nine German tourists, an Egyptian bus driver, *and wounding* eight others.

The U.S. cruise missile attacks on targets in Afghanistan and the Sudan on August 20, 1998 reflected the fact that U.S. intelligence had reliable information that Osama Bin Laden, a leading sponsor and financier of terrorism, planned large-scale attacks on U.S. targets. The U.S. attack on the Shifa Pharmaceutical Plant in Khartoum was a preemptive attempt to prevent the production and use of VX nerve gas by Bin Laden's organization.

There are many causes of transnational terrorism in the Middle East, and many different Western targets:

- The U.S. is a major target because it projects the most power into the region, because of its close ties to Israel, because attacks on the U.S. produce the most world-wide publicity, and because the U.S. can often be used as a proxy for less popular attacks on Middle Eastern regimes.
- The breakdown in the Arab-Israeli peace process makes both Americans and others a target in Israel and other parts of the Middle East, both out of frustration and in an effort to break up the peace process.
- The failures of Middle Eastern secular governments, state terrorism and authoritarianism, economic hardship, social dislocation, and the alienation of youth combine to create extremist groups that not only attack their governments, but use Western targets as proxies. Motives can include attempting to drive out the Western military forces that provide Middle Eastern countries with security, cripple the economy to weaken governments, or win public recognition in the region. While some of these groups are secular, many claim to be Islamic in character. Some totally reject both secularism and any ties to the West or Western values.
- European nations can become the scene of attacks by opposition groups on the Embassies of Middle Eastern regimes, or by opposition groups attacking each other. Iran has sponsored state terrorist attacks on the People's Mujahideen and Kurdish opposition groups in France, Germany, Switzerland and Turkey. Israel has killed Palestinians in nations like Norway.
- Western tourists and businessmen can be the targets of terrorists in the Middle East, as such groups seek to put economic pressure on local regimes, or prove their status and power. For example, an Algerian terrorist group called the GIA (Armed Islamic Group) killed seven foreigners in Algeria in 1997, bringing the total number of foreigners the GIA has killed in Algeria to 133 (since 1992). Bombs have been used in civilian areas in Bahrain, although Westerners have not been major targets. Four U.S., employees of Union Texas Petroleum, and their Pakistani driver were shot and killed in Karachi on November 12, 1998, when the vehicle they were riding in was attacked by

terrorists that seem to have been affiliated with Middle Eastern extremist groups.

- The West can be attacked on the basis of its values, and for corrupting Islamic countries and supporting secular regimes. While the U.S. is the primary target of such attacks, figures like the Saudi terrorist financier Osama Bin Laden want to drive the West out of the region. Bin Laden has previously claimed responsibility for anti-U.S. attacks in Somalia and Yemen. These include an attempt to bomb 10 U.S. servicemen in Yemen in December, 1992. Bin Laden supplied arms to extremist groups opposing Operation Restore Hope in Somalia in 1993, and seems to have financed the groups responsible for the World Trade Center bombing in 1993. He may have helped to finance and organize the bombing of the National Guard headquarters in Riyadh. He threatened to attack U.S. forces in Saudi Arabia to force a U.S. withdrawal from the region in March 1997. He continued to make statements threatening Western interests throughout 1997 and 1998, and made threats that identify him as a possible sponsor of the bombings in Kenya and Tanzania. Bin Laden is claimed to finance and "guide" Arab "Afghani" movements with up to 3,000 members that have conducted operations in Somalia, Chechnya, Afghanistan, Bosnia, Tajikistan, and Yemen.

The most serious challenge the West may face from the Middle East may be the risk that proliferation interacts with terrorism. At present, this is only a possibility. No Middle East state or faction has yet made any attempt to use weapons of mass destruction in transregional attacks on the West, although the U.S. launched cruise missiles strikes against the Shifa Pharmaceutical Plant in Khartoum on August 20, 1998, in a preemptive attempt to prevent the production and use of VX nerve gas by a leading terrorist – Osama Bin Laden.

The basic problem for the West, and indeed for the Middle Eastern states which are as likely to be terrorist targets, is that terrorist attacks using weapons of mass destruction present a fundamentally different kind of threat. They are a far more lethal kind of terrorist threat than the West has yet faced, and they offer radical Middle Eastern states and terrorist factions a relatively easy way to bypass many of the defenses proposed as part of counterproliferation.

Under many conditions, a single act of terrorism can kill thousands of people and/or induce levels of panic and political reaction that governments cannot easily deal with. Under some conditions, the use of weapons of mass destruction can pose an existential threat to the existing social and political structure of a small country — particularly one where much of the population and governing elite is concentrated in a single urban area.

Western Power Projection

Gulf and Middle Eastern security is critically dependent on de facto alliance between the moderate states in the Middle East and the West, and their access to help from western power projection capabilities. The U.S. is only funding about $40-45 billion out of the $60-75 billion in procurement funds necessary to achieve its force goals and the "evolution in military affairs" (EMA). Readiness is down, and the overall force is so badly over-deployed that major retention problems now exist. The high profile of the U.S. in the Gulf has also caused sufficient political backlash *that* the U.S. is *now* a major target for terrorists and has *had* to reduce its profile in the region.

This, unfortunately, is the "good news". NATO European forces are being cut more quickly than U.S. forces and suffer from much more serious underinvestment. NATO has never developed a collective approach to power projection in the MENA region and is unlikely to do so. Europe is deeply divided over the attention and role it should play in dealing with the problems in Algeria and North Africa. Britain and France are the only NATO powers capable of meaningful power projection to the Gulf, *but* they have minimal strategic lift, and only about half the potential pool of forces they had in 1990. Egypt and Syria have tenuous capabilities and little current willingness to use them.

Western power projection will be made steadily more complicated by proliferation and the development of more dangerous forms of asymmetric warfare. Neither the U.S., its Western allies, or its allies in the region as yet have a clear counterproliferation option or surplus funds to pay for such an option.

Geoeconomic Uncertainties: Chaos Theory and the Asian Demand Problems

The internal geopolitical uncertainties in the Middle East are only part of the story. There are broader global geoeconomic problems which can have a major geopolitical impact on the Middle East. While each of these problems involves uncertainties in the political and economic trends outside the Middle East that cannot be analyzed in depth in this study, it is important to understand that the geopolitics of energy in the Middle East are inevitably determined as much by external factors as by what happens in the region. Like it or not, there is no meaningful way to predict the future. The impact of complexity or chaos theory is critical and unavoidable, and efforts at prophecy that ignore these realities depend upon sheer luck for their validity.

One key problem is the growth in global demand for oil imports outside the Middle East. Most recent forecasts of future demand for oil have assumed that the developing nations of Asia will experience very high average sustained economic growth rates over the period through 2020, and that this will create a massive new demand for energy and oil imports.

The collapse of many Asian economies during 1997 has ensured much lower growth in the near term. The Asian economic melt-down was a key factor that led to the massive "crash" in the global demand for oil that began in late 1997, and which caused a sudden 35%-40% cut in average oil revenues in the Middle East. It also affected the investment climate both in terms of the resources of local states, and the willingness of foreign investors to make discretionary investments.

It is too soon to predict the length of the recession/depression in Asia and its overall impact on the economies and investment patterns in the Middle East, although some important signs of recovery had emerged by mid-1999. Even so, it seems unlikely that Asian *oil* demand will recover to even its pre-crash level until sometime beginning in 2001 at the earliest.

The timing and scale of the Asian economic recovery, and the resulting rate of increase in the demand for oil, will shape much of the geopolitics of energy. In 1995, Asia imported about 8.8 million barrels *of oil* a day from the entire Middle East, 8.7 million barrels *of which* came from the Gulf. This was about 50% of total Middle Eastern exports of 17.7 million barrels a day, and 56% of total Gulf exports of 15.4 million barrels a day. In 2020, the Department of Energy projects that Asia will import about 20.3 million barrels a day of oil from the entire Middle East, and 20.1 million barrels a day from the Gulf. This would be about 54.5 % of projected total Middle Eastern exports of 39.8 million barrels a day, and 54% of total Gulf exports of 37.2 million barrels a day.

This rise in Asian dependence on Middle Eastern and world oil exports may occur, but it is scarcely a certainty. It also has two corollaries. First, it raises questions about how long the U.S. will pay to provide all of the power projection to defend the flow of Asian oil. Second, it means that up to 20.1 million barrels of oil per day must flow out of the Gulf to Asia without an interruption or delay. This is an increase of 11.4 million barrels a day from a total of 8.7 million barrels a day in 1995. While the Department of Energy may be exaggerating Asian growth between now and 2020, recessions do not last and Asia's demand could rise for other reasons like the failure of nuclear power programs and problems with coal, nuclear energy, and the environment.

Geoeconomic Uncertainties

The role of competing suppliers presents another geopolitical problem where it is impossible to make clear predictions, but where external factors could have a major impact on the geopolitics and geoeconomics of Middle Eastern energy. Current forecasts generally assume that the FSU, West Africa, Latin America, and the Caribbean Basin (which includes Mexico) will all steadily make major increases in their production capacity, respond efficiently and consistently to market forces, and be major exporters. There are strong economic incentives for virtually all of the exporting nations in these regions to maximize annual income, and this seems to be the most likely case.

Political, economic, and military disruptions, however, can occur outside the Middle East, however, as well as within it. There already are serious stability problems in the FSU states, Nigeria, Angola, Columbia, and Venezuela. Any prolonged underinvestment in energy development in these region, or interruption in production and exports, would rapidly change the demand for Middle Eastern oil. At the same time, current finding rates indicate that West Africa has the potential to increase exports to much higher levels than are currently projected in the EIA and IEA models.

It is important to stress that the political and economic stability problems in the FSU states, Nigeria, Angola, Columbia, and Venezuela create strong pressures to maximize oil and gas export revenues, and restrict domestic demand. As a result, they can have a *stabilizing* effect on the world economy in the sense that they tend to ensure high levels of production and exports and to limit prices.

The economic impact of a temporary cut in exports from any given country would also be relatively limited, even if the world economy showed high rates of economic growth and increases in the demand for oil, because of the elasticities between energy consumption and growth, and the ability of other regions to provide additional exports at moderate increases in price. Such cuts would normally simply cause several weeks or months of adjustment in selected markets and act as a windfall that increased revenues to other exporters.

The problem raised by interruptions in energy supply from other regions would only become critical in global geoeconomic terms if other interruptions in oil supply occurred simultaneously. Such a scenario might take the form of interruptions in the exports from several different countries at the same time, or a case like an interruption from a critical supplier like Russia that halted most or all oil and gas exports for a prolonged period.

Similar problems could also *happen* if an interruption in oil and gas exports occurs at the same time as a crisis in other aspects of supply like nuclear or coal, or if a structural problem occurs like prolonged underinvestment in energy production by a critical supplier like Russia. Once again, it is critical to understand that efforts to structure or predict scenarios for such cases are exercises in chaos theory. It is possible to warn that they can occur. In fact, it is a historical paradox that history is shaped largely by the changes caused by combinations of unpredictable events rather than the most likely combination of predictable events. This is a critical issue in both understanding the realities of Middle Eastern energy exports and the nature of global energy stability.

It is also important to note that geopolitics does not center around global energy supply or prices. Markets dependent on a given supplier will often be acutely sensitive to energy interruptions that are limited in scale. Europe, for example, is highly dependent on Russian gas. Kuwait was not critical to the world's energy supply and Iraq did not rush to cut exports. The importance of the stability of Venezuela is not measured simply in MMBD. The political and strategic content of energy interruptions will often be more important than the economic context, and this makes the risk of any given interruption much harder to estimate. It also creates the risk that the political and military tensions that cause an energy problem outside the Middle East could interact with the tensions inside the region.

Regional Energy Issues

Regional geopolitics also interact with practical problems in maintaining an expanding supply of energy within the Middle East. Much of the economy and security of the Middle East is shaped by its energy reserves, energy exports, and the revenues it receives from the sale of oil and gas exports. Given the present oil crash, the most serious energy issues affecting the region are the impact of low oil revenues on economies which have failed to modernize and diversify, the impact of sanctions on several critical suppliers, the size of future demand for exports, national policies to increase production and export capacity, and the ability to obtain the investment necessary to implement those policies.

At the same time, the short-term problems that low oil and gas revenues and sanctions are creating for Middle Eastern governments are not new problems. They are simply exacerbating long-term structural problems in the regional oil-exporting economies that are caused by a number of economic and demographic pressures. There is nothing new about major swings in oil revenues, and there have been vast swings in oil revenues in the past. OPEC oil revenues were worth around $77 billion in constant 1990 dollars in 1972. After the October War and the 1974 oil embargo, they leapt to levels of around $340 billion and then dropped back to less than $300 billion during 1975-1978. The fall of the Shah of Iran and the start of the Iran-Iraq War drove them to a new peak in 1980, when they were worth $489 billion. An oil price collapse began in 1985, and revenues dropped to $83 billion in 1986. They gradually rose back to levels of around $150 billion a year in early 1997, but a new "oil crash" began late that year that still continues. Total OPEC revenues were only $80 billion in 1998, and are estimated to rise to $107.7 billion in 1999.

The nations of the Middle East also suffer from very high rates of population growth, the inability to sustain past welfare and entitlement programs, and a lack of economic diversification. This is made worse by excessive spending on show piece state-sponsored industrial and petrochemical projects, excessive military spending, a weak private sector and high outflows of capital to other regions, and a services sector that stresses imports or import-dependent light manufactures with little value added.

Given this reality, the central energy issue in the region is the need to create new economic structures that will develop energy exports while achieving internal stability by offering suitable employment and incentives for investment. There is a growing feeling that the future increases in demand and production may be significantly lower than many experts forecast a year ago. The current crisis in oil revenues has also made it clear than many countries cannot hope to fund either their domestic infrastructure needs or the development of their oil and gas resources without substantial foreign and private domestic investment. Many countries in the region are beginning to face this fact. They are beginning to rethink their plans to increase production capacity, *in favor of* private and foreign investment.

Uncertainties Regarding Oil and Gas Production and Reserves

The previous analysis has shown that there are major uncertainties relating to world demand and the availability of energy from other sources and regions. There are also sharply different estimates of the oil and gas that Middle Eastern states can produce and of the energy reserves available.

Uncertainties in Estimates of Oil Production

Many oil companies and private experts *contend* that the EIA and IEA estimates of future demand and Middle Eastern production are over-optimistic and are based on demand-driven models that do not examine the real-world constraints in increasing supply. At the same time, no company publishes its own estimates in detailed enough form to be useful.

One of the issues that emerges from these figures and tables that affects all aspects of energy policy and planning is the need for a broader consensus as to how to model and estimate the future trends in energy supply and demand. The International Energy Agency and the Energy Information Agency used what might be termed as "demand-side" models where the estimates of oil and gas production and exports are driven largely by assumptions about economic growth, and production is assumed to meet the rise in demand. These models have inevitable uncertainties. It is almost impossible to forecast world economic growth, the future price elasticities for every form of energy, the impact of environmental issues, and the willingness and ability of given energy suppliers and exporters to provide a given type of energy at a given time at a given price.

At the same time, the "supply-side" models used by the oil and gas industry, and by other energy suppliers, consist largely of a group of scattered estimates by type or category of energy that cannot be coordinated into a meaningful picture of global and regional trends. In many cases, such models tend to be too conservative because industries are very sensitive to short-term risk. If demand-supply side models tend to exaggerate Middle Eastern energy production because they take an over-optimistic view of sustained world economic growth, supply-side models can be too conservative.

The problem is made worse when the data, assumptions, and modeling techniques are not fully disclosed. This often makes it impossible to know how different estimates really are, and to distinguish back of the envelope guesses from serious modeling efforts. It also makes it impossible to reconcile the differences affecting the estimates of Middle Eastern energy production between macro-economic demand-side models and supply-side estimates of the oil industry. It is equally difficult to create a suitable range of scenarios for emergency planning and for analyzing the flow of oil and gas exports.

The EIA and IEA models can certainly be brought more up-to-date in any given case, but using the latest data and assumptions does not mean that the end result is any less uncertain or that that results are likely to prove any less ephemeral over the mid to long-term. The fact remains that models can illustrate possible futures, but cannot predict the future over any period of time in which large-scale discontinuities suddenly occur in existing trends. This, unfortunately, includes the entire record of recorded history.There are, however, two policy-level conclusions that can be drawn from this analysis. One is that it is as dangerous for governments and policy makers to rely on demand-driven models like the ones used by EIA and the IEA as it is for industry analysts to rely on short-term intuitive estimates based on supply-driven considerations.

When uncertainty cannot be eliminated, it must be planned for. This does not mean investing in *possible* solutions to worst cases – an approach that must inevitably waste vast amounts of public funds in trying to deal with unlikely contingencies. In the case of Middle Eastern oil and gas production, however, it does mean that high levels of sustained investment in increased production capacity will act as a stabilizing influence on world energy supply and demand. This stabilizing impact will minimize the risk of sudden oil shocks, as well as limit the impact of problems in other forms of energy supply. It will also probably maximize the income of exporting nations over time because it creates a stable and predictable market and climate for investment.

Uncertainties in Estimates of Oil Reserves

The size of commercially exploitable Middle East oil reserves is as critical an issue as the rate of increase in production capacity and exports. Saudi Arabia alone contains 261.5 billion barrels of proven oil reserves (more than one-fourth of the world total) and up to 1 trillion barrels of oil in place. The Saudi-Kuwaiti Neutral Zone contains about 5 billion barrels of proven oil reserves. More generally, conventional wisdom indicates that nearly two-thirds of the entire world's proven reserves are located in the Middle East.

It seems doubtful that these issues regarding the size of Middle Eastern and Gulf reserves will have much strategic impact until well after the year 2010. Nevertheless, they are a warning about an over-simplistic approach to calculating reserves and production figures, and transforming the figures into estimates of long-term strategic importance. They also are a warning about the difficulty of estimating how long reserves will last, recovery capability, and cost.

Uncertainties in Estimates of Gas Production

It is considerably *more difficult* to illustrate the uncertainties relating to Middle Eastern gas production and reserves than it is to illustrate these uncertainties for oil. Gas is only beginning to emerge as a major export, market trends are harder to predict, and the modeling effort is much less sophisticated. At present, gas liquids (gas oil) and LNG are also sufficiently high priced so that they go to niche markets, and normal gas shipments are restricted to pipelines. As a result, the Middle Eastern use of gas as a substitute for the domestic use of oil and as a feedstock for petrochemical related exports has considerably more impact on the global economy than gas exports.

Uncertainties in Estimates of Gas Reserves

There are additional problems in estimating gas reserves, compounded by the fact that gas is still an emerging fuel in the Middle East, and that exploration has been limited in key countries like Iran, Iraq, and Saudi Arabia. The Middle East plays a critical role in terms of both world reserves and export potential, and that Iran dominates proven gas reserves in the Middle East.

It is important to note, however, that such estimates are speculative, and that major debates exist over the true size of gas reserves, the cost and potential market for gas, the size of the domestic market, and the gas policies that given nations should pursue. Similar uncertainties affect the gas reserves and policies of many other Middle Eastern states, as well as those of the Caspian and Central Asian states that might ship or swap oil in ways impacting the Gulf.

Impact of Low Oil and Gas Revenues

The short-term crisis in Middle Eastern oil revenues is largely a function of the Asian economic melt-down, and similar problems in Russia and Latin America. The long-term problems, however, are largely a function of structural economic problems and pressures which long precede the current crisis. The oil-exporting Middle Eastern states have consistently under-performed the non-oil exporting states in the region, and have fallen below Sub-Saharan Africa in terms of economic growth.

As a result, there has been a series of unexpected cuts in average oil revenues *approaching* 30-40%. These cuts have *hindered* the region's ability to maintain *both its* welfare payments and entitlements, and short-term investment. It is too soon to predict the length and intensity of these cuts in oil revenues, but they will, at a minimum, have a serious short-term impact *on* every major oil and gas producer in the Middle East.

The current economic crisis has not created any of the major structural economic problems that may ultimately affect the geopolitics of gas and oil. It has, however, been a major catalyst in showing that low oil prices and low oil revenues can create major problems for exporting countries. Its impact is illustrated by the fact that Asia imported about 8.8 million barrels of oil a day from the entire Middle East in 1995, 8.7 million barrels of which came from the Gulf. This was about 50% of total Middle Eastern exports of 17.7 million barrels a day, and 56% of total Gulf exports of 15.4 million barrels a day. These import levels continued to rise at moderate price levels until the Thai economic crisis in the fall of 1997. The Thai collapse then triggered a general collapse of most Asian economies that spilled over into other parts of the world and led to major cuts in the demand for Middle Eastern oil exports and a serious drop in oil prices.

The Decline in World Oil Prices

The Department of Energy estimates that decline in world oil prices has had serious implications for OPEC oil export revenues, balance of payments, budgets, and overall economic conditions. Since late 1997, the OPEC "basket" price (a weighted average of Algeria's Saharan

Blend, Indonesia's Minas, Nigeria's Bonny Light, Saudi Arabia's Arabian Light, Dubai, Venezuela's Tia Juana, and Mexico's Isthmus) has fallen more than $7 per barrel (around one-third). This is a drop from nearly $19 per barrel in October 1997 to $11.91 per barrel in August 1998. In constant 1990 dollars, world oil prices during 1998 have been running at the lowest levels since 1973, prior to the Arab Oil Embargo late that year, and since 1986, following the oil price collapse of late 1985/early 1986.

Reverse Oil Price Shock and the Impact of Low Oil Export Revenues

The end result was what the Department of Energy called a "reverse oil price shock," in which prices moved sharply downward in contrast to the sharp oil price increases of 1973/4, 1979, and 1990. Low oil prices are causing a "shock" to oil exporting countries, while the economies of oil consuming countries are being positively impacted. In fact, the current situation was similar to, but not as extreme as, the oil price collapse of 1985/6, when oil prices were cut in half.

The Department of Energy estimated that these low oil prices were caused by four main factors, including:

- OPEC's December 1, 1997 agreement to raise the group's production quota;
- A warmer-than-normal winter (1997/1998) in the northern hemisphere;
- Increasing Iraqi oil exports; and
- Reduced oil demand due to the severe economic crisis in East Asia).

Prices recovered somewhat in 1999, but only at the cost of cuts in exports, leaving oil revenues relatively low. If low oil revenues continue for several years (and no obvious end to low oil revenues appears in sight), the resulting cut in government revenues will probably force many Middle Eastern countries to cut their budgets and development plans in ways that result in significant economic, social, and political tradeoffs. The International Monetary Fund stated in May 1998, *that* the decline in oil export revenues "would pose a serious risk to the growth outlook " for the Persian Gulf region, "and particularly for the region's largest oil exporters such as Saudi Arabia and Kuwait ...if sustained." Current indications make it probable that many countries will experience negative real growth, particularly in their GNP. It is already clear that the situation is serious. While there currently *are* no reliable figures on the impact of the "oil crash" on the Middle East as a region, OPEC is dominated by Middle Eastern states and the trends in OPEC provide a broad picture of what is happening in the Middle East. The OPEC Basket Price for oil averaged around $12.37 per barrel during the first 6 months of 1998, compared to $19.30 during the first six months of 1997. During the first 6 months of 1998, OPEC countries earned $53 billion in revenues from the export of crude oil. This represents a direct reduction in OPEC crude oil export revenues of $22 billion (30%) from the first six months of 1997. For 1998 as a whole, OPEC earned slightly less than $100 billion in oil export revenues, down at least 35% from $154 billion in 1997.

Whatever happens, even sustained low oil revenues would only exacerbate a longer-term trend. In real terms (constant 1990 dollars), OPEC revenue peaked in 1980, at $439 billion. OPEC's worst revenue year in constant dollar terms since the early 1970s was in 1986, following the oil price collapse of late 1985/early 1986. In 1986, OPEC earned $83 billion (in constant 1990 dollars), compared to $77 billion in 1972. OPEC revenues for all of 1998 were $80 billion (in constant 1990 dollars). OPEC revenues in real terms in 1998 were less than one-fifth 1980 revenues. To put this in perspective, 1998 was the worst year for OPEC revenues (in real terms) since 1972 including 1986, following the oil price collapse of late 1985/early 1986. *There are also few indications that oil revenues will rise sharply in the near future.*

The resulting economic pressures will vary by country. Countries like Algeria and Iran, for instance, contain relatively large populations and relatively small oil reserves. These countries have tended to favor a strategy of short-term revenue maximization, and have relatively low political/social tolerance for the pain caused by low oil revenues. Countries with smaller populations and large oil reserves, like Kuwait, the United Arab Emirates, and Saudi Arabia, on the other hand, have tended (also with exceptions) to favor a strategy of long-term revenue maximization, and generally have been in stronger positions to weather price declines. This situation may be changing, however, since Kuwait, the United Arab Emirates, and Saudi Arabia now are experiencing major budget deficits and have just as much incentive to maximize short-term revenues as long-term revenues.

The effect of oil price fluctuations on oil producing countries also varies according to whether a given country is a relatively low-cost or high-cost oil producer. For low-cost producers (like Saudi Arabia, Kuwait, the United Arab Emirates, and Iraq), the marginal cost of producing each additional barrel of oil is relatively low, and oil production usually remains economical when oil prices fall. For relatively high-cost oil producers like the United States, Norway, the United Kingdom, or Canada, on the other hand, low oil prices can turn many oil fields from economical to uneconomical in a short period of time or limit investment in the development of existing fields and production from new ones. High oil prices have generally tended to encourage oil exploration and production in previously marginal fields (like the North Sea in the late 1970s and early 1980s). Virtually all Middle Eastern production is low cost by the standards of other regions.

The OPEC Quota Game

One side effect of these trends has been new efforts by producers to agree on cutbacks in production as a way of raising prices. This has led both to the revival of OPEC quotas and to new political arrangements including nations like Mexico. As will become clear from the country-by-county analysis that follows, however, most Middle Eastern countries have long either violated their OPEC quotas or played semantic games that have largely vitiated them. The fact Iraq now has virtually open-ended quota and license to export under the oil-for food program also limits the value of any such agreements.

One key factor that limited cooperation by OPEC and non-OPEC exporters was the fact that many were so desperate to maximize short-term revenue and saw little prospect that revenues would rise significantly if they strictly adhered to any quotas. Another problem was that OPEC is the only formal body for such cooperation, but only represented less than half of world production. OPEC's share of world oil supply has remained constant at about 39.5% between 1993 and 1996, while total output from non-OPEC nations, especially the North Sea, Latin America, and West Africa, has increased sharply. OPEC oil production did increase by about 1.6 million barrels per day (b/d) in 1997, however, and by another 0.6 million barrels per day in 1998, out of total global oil supply growth of 3.2 million b/d between 1996 and 1998. This increased its share of the world's oil supply by more than one percentage point, although scarcely enough to increase its power as a cartel. The EIA estimates that the members of OPEC accounted for 40.6% of total world oil supply in 1998.

OPEC reacted to these problems by holding another ministerial meeting on November 25 and 26, 1998. Some countries like Algeria and Kuwait had expressed hopes that further production cuts might be agreed upon, but OPEC did not extend the current production levels, nor did it cut oil production even further. In fact, arguments between Iran, Saudi Arabia, and Venezuela effectively paralyzed collective action. As a result, OPEC will meet again in Vienna, Austria on March 23, 1999 to determine whether more oil production cuts can be agreed upon. The EIA estimates that the lack of concrete steps taken by OPEC on oil production targets was a major factor in the OPEC basket price dropping below $10 per barrel ($9.89) on November 30, 1998.

This price "crash" finally pushed both OPEC and non-OPEC exporters into more effective action. In March 1999, major OPEC members reached a preliminary agreement to cut world oil production by more than 2 million barrels per day beginning on April 1st. Special emphasis was placed on resolving the dispute over Iran's adherence to this and previous output cut agreements. Iran had maintained that its 300,000 bpd reduction should be taken from a baseline of 3.9 million bpd, whereas other OPEC states said it should be 3.6 million bpd. Saudi Arabia, Kuwait, and the UAE agreed to absorb the missing Iranian cut and Iran has accepted the benchmark of 3.6 million bpd for all future cutbacks in return. All other OPEC states will be cutting production from their July quotas. Every OPEC member except Iraq agreed to the cuts.

Iraq was an dis an important "wild card." Iraqi production has previously rendered some OPEC cutbacks ineffective. Iraqi production increases cancelled out most of the benefits of cuts made in April 1998. While OPEC output fell by nearly 1 million bpd, Iraqi output rose by more than 400,000 bpd, thus effectively reducing the OPEC cutback to only 500,000 bpd. The tensions between Iraq and the United States over weapons inspections do create the continuing possibility that the Iraqi oil-for-food deal and Iraqu exports could be interrupted. If Iraq were to stop exporting, some 2 million bpd would be taken off the market. It seems more likely, however, Iraqi production will continue and increase. An Iraqi export figure of 2 million bpd is now viewed almost as a constant. Iraq is also entitled to export larger volumes within the oil-for-food deal, although it now lacks the capacity and facilities needed to increase production.

Before OPEC's March meeting, Iraq's Oil Minster Amir Muhammed Rasheed stated that "Iraq has submitted a specific plan to OPEC ministerial meeting…to cut production by 2.5 million bpd and this should be borne entirely by Saudi Arabia" in order to raise oil prices. Rasheed accused Saudi Arabia of flooding oil markets with its production in order to suit an U.S. policy that aimed to control oil prices. He said that Arab countries had lost $929 billion between 1987 and 1998 because of the Saudi oil policy.

It seems doubtful that anything like an energy cartel will emerge in the future, or alter the broader trends in production just discussed. More broadly, it is not clear that any action that OPEC and other producers can take will make enough difference in terms of marginal production and price levels to have a serious impact on market forces or a serious negative impact on consumers. If anything, the short term impact of any rise in prices might be offset by the mid to long-term advantages in terms of the resulting political and economic stability of exporting states and increased investment in production capacity.

Near Term Prospects

The EIA estimates that there is only a moderate probability of a rapid improvement in oil revenues. It estimates that downside risks (i.e., the chances that revenues could decrease even further) are strong. Factors that could help push oil revenues even lower include: a renewal and/or broadening (i.e., to China and/or other countries) of the East Asia crisis, which could reduce marginal world oil demand still further; a failure by OPEC to cut as much production as it actually has pledged; a sharp downturn in Russia's economy, which could reduce Russian oil demand and put more Russian oil on world markets; and continued increases in Iraqi oil output.

The "upside" risks (i.e., the chances that revenues could rebound) include: a more rapid recovery in Asia than expected; a disruption in Iraqi oil exports; and a colder-than-normal winter in the northern hemisphere.

Due to increased compliance with OPEC production cuts, oil prices rebounded in the first quarter of 1999. OPEC basket prices reached a 1999 low of $9.96 in February, but quickly rose to $15.43 in May. The New York Times reported on June 7, 1999, that crude oil for delivery in August rose to $19.78, the highest price since November 1997. This price increase coincided with an increase in compliance with OPEC production cuts, which according to the IEA, reached 91% (excluding Iraq) in June 1999.

The increase in oil prices that began in April 1999, however, should not be considered a panacea. In this case, higher oil prices will not necessarily result in higher revenues since less oil is being sold. It is also important to note that most Middle Eastern countries face serious revenue and budget problems even when oil sold for $17 a barrel in 1997 dollars. As a result, the various efforts to raise prices and revenues by cutting production will have only a marginal short-term impact even though they have so far been successful. Middle East exporters will take years to recover from the current shock if oil prices do not rise above $20 a barrel in the near future, and most countries would require sustained periods of $25-$30 oil to balance their budgets and implement the investment plans they have developed before the current oil shock. Even limited rises in oil prices can greatly ease the current oil shock in terms of short term budget crises, debt rescheduling, etc., but only major structural reform can do more than delay a steady increase in economic, political, and social problems.

Impact of Sanctions

UN and U.S. sanctions affect three critical producers in the Middle East: Iran, Iraq, and Libya. There is no way to know exactly how long these sanctions will last, or how much impact they will have on each country's energy development. It is clear, however, that sanctions have had an impact on short-term energy developments in three important Middle Eastern oil and gas exporting states. The EIA currently projects that several of these states will make major increases in their oil production capacity and they may also make major increases in gas production.

The EIA's reference case projections of present and future oil production are shown in Table 1 below, and in the figures at the end of this section.

Table 1. EIA Estimates of the Future Oil Production of Sanctioned Middle Eastern States

	1990	*1997*	*2000*	*2005*	*2010*	*2015*	*2020*
Iran	3.2	3.9	4.0	4.3	4.5	4.9	5.5
Iraq	2.2	1.6	2.8	3.2	3.8	4.7	5.9
Libya	1.5	1.5	1.5	1.6	1.7	1.6	1.5
Total "Rogue"	6.9	7.0	8.3	9.1	10.0	11.2	12.9
Total Gulf	18.7	22.8	23.9	28.1	29.6	34.9	42.2
Total OPEC	27.2	33.0	34.8	40.7	43.3	48.7	55.9

It is unclear that Iran, Iran, or Libya can meet theses goals in the short-term because of sanctions. Each country is continuing to invest in energy development, but there is no question that sanctions have meant serious cutbacks in the scale of investment, and have greatly complicated energy planning. The impact of sanctions also interacts with the current drop in demand for oil imports and in oil revenues – which create additional reasons to avoid or delay investment.

Sanctions can present significant problems in developing the most cost-effective expansion of oil and gas production in several critical countries, and could increase the vulnerability of world oil markets if supply interruptions take place in other areas. Sanctions against Iran also complicate the cost-effective development of Caspian and Central Asian oil resources, while U.S.-driven sanctions tend to exclude U.S. companies from the Iranian, Iraqi, and Libyan markets without imposing similar constraints on their competitors.

More broadly, sanctions against energy investments ignore the fact that such investments actually require counterpart investments from the sanctioned country, and that the cash flow from abroad is tied to progress payments over a period of years. They ignore the fact that the profits from many investments only come three to five years in the future, when the government may well have changed.

There is, however, another side to this argument. The recent *decline* in the growth of the demand for oil and gas exports have eased the pressure for increases in Iranian, Iraqi, and Libyan production, and it can be argued that sanctions now have the short-term effect of transferring export revenues to moderate and friendly states. The "oil for food " controls on Iraqi imports are a key way of preventing imports of arms and dual-use equipment and are an increasingly important substitute for effective IAEA and UNSCOM inspections. The fact also remains that Iranian arms imports are now about one-third of their level during the Iran-Iraq War, Iraq has had no arms imports since 1990, and Libyan arms imports are a fraction of their pre-sanction level. The arguments for and against sanctions are not simply a matter of energy policy, and are considerably harder to resolve when security issues are considered as well as economic and energy factors.

Sanctions Against Iran

The key legislative sanction affecting energy development in Iran is the Iran-Libya Sanctions Act (ILSA), which was passed unanimously by Congress and signed into law by President Clinton in August 1996. ILSA imposes mandatory and discretionary sanctions on non-U.S. companies which invest more than $20 million annually (lowered in August 1997 from $40 million) in the Iranian oil and gas sectors.

The passage of ILSA was not the first U.S. sanction against Iran. Long-standing legislation prohibits U.S. firms from exporting arms and dual-use technology to Iran, and imposes sanctions against foreign countries and companies that do so. In early 1995, President Clinton signed two Executive Orders that prohibited U.S. companies and their foreign subsidiaries from conducting business with Iran. The Orders also banned any "contract for the financing of the development of petroleum resources located in Iran. " One key result of these orders was that U.S.-based Conoco was obligated to abrogate a $550-million contract to develop Iran's offshore Sirri A and E oil and gas fields.

On August 19, 1997, President Clinton signed Executive Order 13059 reaffirming that virtually all trade and investment activities by U.S. citizens in Iran are prohibited. These Executive Orders blocked what seems to have been a major initiative by then President Rafsanjani to improve relations with the U.S.. Combined with ILSA, they are a major barrier to improving relations with the more moderate regime of President Khatami.

It is unclear how effective these sanctions have been, and it is impossible to separate the impact of sanctions from Iran's poor management of its economy and foreign investment programs that have probably done more to hurt Iran's energy development than sanctions. *Regardless*, the threat of secondary U.S. sanctions has deterred some multinationals from investing in Iran. In August 1996, Australia's BHP pulled out of a proposed $3-billion pipeline project to transport Iranian natural gas to Pakistan and India. *In addition, U.S. Secretary of State Madeleine K. Albright noted that U.S. efforts to discourage the Indonesian firm Bakrie from proceeding with the development of the Balai oilfield contributed to Bakrie's apparent decision to withdraw, although the impact of the Asian financial crisis was also important.*

Since that time, however, other firms have committed to sizable oil and gas sector projects in Iran in spite of the U.S. sanctions. Following the passage of ILSA, Iran and Turkey signed a 22-year gas supply deal worth an estimated $20 billion. Despite U.S. government opposition, Turkish officials stated in March 1997 that they would proceed with the construction of the required infrastructure.

France's Total and Malaysia's Petronas are proceeding with development of the Sirri A and E oil and gas fields, a project that has been underway since 1995. They are expected to increase production by 120,000 b/d by mid-1999. A Total-led consortium will help develop the giant South Pars gas field, and Total has stated that it is interested in leading a consortium to create a new pipeline through Iran that could be used to export oil from the Caspian and Central Asia.

The U.S. has faced major opposition to its sanctions on foreign countries from the European Union, and less openly from major trading partners like Japan. No other country in the world formally supports or endorse U.S. sanctions, *but* virtually every major ally in the world except Israel has criticized this *particular policy*. As a result, the U.S. tends to "speak stickly and carry a big soft. " It has increasingly tried to find reasons to grant waivers for such deals, although it still seeks to ensure that pipelines do not go through Iran from the Caspian and Central Asia.

In the spring of 1998, the U.S. gave Total a waiver on the grounds that its Sirri deal, despite the $600 million size of *the* investment, was signed prior to ILSA's August 1996 enactment. Petronas, which acquired a 30% stake in the Sirri deal in 1996, stated in early March 1998 that it would not withdraw from the project despite U.S. objections. This, to all practical purposes, has been interpreted as U.S. recognition that it would be forced to grant similar waivers in the future.

Sanctions have given Iran little trouble in gathering foreign bid packages for its oil and gas projects since that time, but low oil revenues, uncertain Iranian politics, and problems in Iranian law *have hindered attempts to increase foreign investment*. Much more is involved than foreign resistance to ILSA. The election of President Khatami has given the U.S. more reason to be tolerant of deals with Iran, and attempts by the United States to implement ILSA have run into opposition from many foreign governments. The European Union has taken a strong stand, *stating* that ILSA is "unacceptable" and has barred European companies from complying with it. As a result, President Clinton has used a provision of ILSA which allows waiver of sanctions for countries that take "significant steps, including the imposition of economic sanctions" towards Iran. In practice he has accepted the EU argument that it has taken significant steps towards restricting Iran's access to weapons of mass destruction. He has also reacted to the EU's warning that imposition of sanctions on Total could lead to a trade war with the U.S..

Most foreign firms now consider a waiver to be a high probability, and the U.S. has taken further steps that reinforce this feeling. In November 1998, the U.S. government relaxed the reporting requirements for foreign affiliates of U.S. companies that engage in petrochemical transactions with Iran. A U.S. State Department official emphasized that *this* was not an attempt by the Clinton administration to improve relations with Iran: "This is not intended as a signal to Iran. " However, the U.S. took the action in the context that Iran is the Middle East's second-largest petrochemical producer after *Saudi Arabia*, and plans to triple its annual output to 30 million tons in 25 years with a $24 billion development program. Iran has plans *to* build about a dozen petrochemical facilities to produce methanol, acetic acid and fertilizers by 2001, and exported $560 million in petrochemicals during the last Iranian year, which ended on March 20, 1998. *That* amount was up from $506 million in the previous year, but short of a $650 million target.

As a result of the U.S. policy change, transactions involving Iranian-origin petrochemicals no longer have to be reported to the U.S. Treasury Department's Office of Foreign Assets Control (OFAC). In addition, affiliates of U.S. companies don't have to report certain goods sold to Iran related to petrochemical transactions, such as oilfield supplies or equipment, or services provided like financing and insurance. The reporting change was requested by a U.S. company that believed the requirement was burdensome.

The U.S. policy change, which was published in the November 10, 1998, Federal Register, did not alter the reporting requirements for crude oil and natural gas transactions. OFAC has required U.S. companies to report certain Iranian oil-related transactions entered into by their foreign subsidiaries since 1995. In November 1996, the reporting requirements were extended beyond transactions in crude oil and natural gas to also cover petrochemicals and certain related goods (oilfield supplies) and services (financing and insurance). Reporting requirements were restricted in April 1997 to those transactions totaling $1 million or more during *a* quarter of a calendar year.

The U.S. policy change also does not help U.S. firms in dealing directly with Iran. The Clinton Administration has waived sanctions on foreign companies, improved relations in some areas, and is seeking to remove Iran from the narcotics list. However, *the Clinton Administration* strongly opposes pipelines through Iran and its executive orders still prohibit U.S. commercial investment and bidding by foreign subsidiaries of U.S. firms. As a result, no U.S. firm bid on 29 oil and gas projects worth several billion dollars which were up for bid on November 30, 1998. Many European and Asian firms bid, however, and Iran seems to have obtained all of the bids it needed.

In fact, no U.S. corporation has actively pursued such investment since the Clinton Administration blocked the Conoco deal although some have bought bid packages from Iran. ARCO, Conoco, and Unocal have not said whether they have purchased bid packages. Mobil has said it has not. The end result, however, is a growing risk that U.S. firms will be excluded at the *time when* Iran most needs foreign investment and technology, and when other countries are willing to meet Iran's needs. This could limit the U.S. commercial presence in Iran long after sanctions are lifted, as well as future sales of pipeline, gas, petrochemical, and enhanced recovery technology.

The U.S. opposition to oil swaps and pipelines though Iran also hurts U.S. industry. Mobil, for example, has sought oil swaps to ship up to 3 MMBD from fields in Turkmenistan that are near Iran for Iranian oil shipped out of Southern Iran. Several U.S. firms could play a major role in selling Iran pipeline technology.

The United States modified its sanctions on April 28, 1999 to allow shipments of donated clothing, food and medicine for humanitarian reasons (trade in informational materials such as books and movies is also allowed). However, all other U.S. sanctions against Iran remain in force. On the same day that the humanitarian exceptions were made, the U.S. denied Mobil's request to swap crude oil from Kazakhstan with Iran.

From a policy viewpoint, U.S. sanctions may prove to be counterproductive in other ways:

- The whole premise behind the U.S. policy seems flawed. Major foreign investment in Iran's energy sector does not give the Iranian government large windfalls. Payments are made only for specific goods and services and on a progress basis. If anything, the money benefits Iran's technocrats and private sector. Any major returns on investment are deferred for years, and Iran must pay a very substantial share out of all foreign buy back deals. In spite of Iran's growing economic problems, the Majlis approved the funding of $5.4 billion in buy back projects for the 1999/2000 fiscal year, the same amount approved for 1998/2000. As a result, investment in energy tends to restrict Iranian investment in arms at least in the mid and near term while benefiting its secular economy, increasing opportunities for U.S. investment, and ensuring adequate, affordable supplies of energy.

- The U.S. has a strong interest in avoiding highly sensitive pipeline politics in the Caspian and Central Asia. It does not wish to either provoke Russia or to see Russia dominate export routes. There are serious questions about the economics and political security of pipelines to Turkey that avoid Iran, and whether oil exports have their natural markets in Europe. Pipelines through Iran present their own problems and complications, but they offer an alternative to Russia and a more direct route to Asian markets. While the U.S. is currently involved in a new "great game" over pipeline politics in the Caspian and Central Asia, it might serve its interests much better by simply relying on market forces. The best way to win this "great game " may be to avoid playing it.
- The U.S. has had no success in compelling Iran to change its policies, *and* has cut itself off from Iran's moderates and strengthened the Iranian hard-liners that charge that the U.S. is the "great Satan." Sanctions may well have produced the exact opposite of their intended effect. Furthermore, they have at least some impact in worsening the economic crisis in Iran, a crisis that may ultimately halt or bring down Khatami and Iran's moderates.
- Iran has clearly demonstrated that it will spend what it takes to proliferate, with or without sanctions. If anything, sanctions strengthen the position of the hard-liners that charge that Iran must proliferate to deal with the threat from the U.S..
- Sanctions have had no impact on the scale of Iran's conventional arms purchases. The latest declassified U.S. intelligence data shows that Iranian arms imports shrank from $3,125 million in constant dollars in 1986, during the peak of the Iran-Iraq War, to 1,800 in 1991. Once Iran saw the scale of the U.S. victory in the Gulf War, and the size of Iraqi losses, its arms deliveries shrank to $931 million in 1992, and then declined to $350 million in 1996 – approximately one-tenth of their level during the mid-1980s when Iran had even more desperate economic problems than it does today. In fact, arms imports have shrunk from a peak of 27.5% of all imports in 1989 to 2.2% in 1996. Even if one only looks at the period after the Gulf War, Iranian new arms agreements dropped from $7.2 billion in 1990-1993, to $1.6

billion in 1994-1997. Ironically, Iranian arms orders and deliveries rose slightly in 1996 and 1997, after sanctions were imposed, rather than dropped.

Sanctions Against Iraq

Sanctions against Iraq present a different problem. Since the end of the 1991 Gulf War with Iraq, UN inspectors have attempted to make Iraq comply with cease-fire terms that require it to destroy its long-range missiles and weapons of mass destruction. Such compliance is a main precondition for lifting of UN sanctions, including the embargo on Iraqi oil exports imposed in August 1990. Iraqi oil exports were then forbidden by UN Security Council Resolution 661 as the direct result of Iraq's invasion and conquest of Kuwait – a state that had been a major ally throughout the Iran-Iraq War.

Sanctions continued after the Gulf War because Iraq never complied with the terms of the cease-fire and remained a major threat to the peace of the region. In April 1995, however, the Security Council passed Resolution 986, which allowed limited Iraqi oil exports for humanitarian and other purposes. Iraq began exporting oil under this resolution in December 1996.

As a result, Iraqi crude oil production (including lease condensates) increased from 781,000 barrels a day in December 1997 to around 2.2-2.3 million in August 1998. Iraqi oil exports also increased steadily, reaching an estimated 1.7 million barrels a day in July 1998 and perhaps as much as 1.8 million barrels a day in August (up nearly 500,000 barrels a day since April 1, 1998). The EIA estimates that this increase in Iraqi oil exports played a significant role in the world oil glut which is responsible for the sharp decline in world oil prices since late 1997.

In spite of further violations of the cease-fire accords, sanctions were again eased to relieve the burden on the Iraqi people. On February 20, 1998, the UN Security Council unanimously approved an increase in the amount of oil Iraq may export, to $5.265 billion over a 180-day period (from $2.14 billion previously). After deducting a predetermined amount for war reparations to Kuwait and to fund operations of UNSCOM, proceeds from Iraqi oil sales are then used to purchase humanitarian goods such as medicine, health supplies, foodstuffs, and materials and supplies for essential Iraqi civilian needs.

Iraq found, however, that it *was only able* to export about $3 billion worth of oil over the initial six-month period because of low prices and technical constraints on production. *As a result, Iraq began to export at its maximum capacity, and export figures of 2 million bpd became almost a constant. Iraq was theoretically entitled to export larger volumes within the oil-for-food deal, but lacked the capacity and facilities needed to increase production.*

It needed major additional infrastructure investments, which are banned under current sanctions, to boost exports beyond this level.

The Security Council renewed Iraq's ability to export for another six months at a level of $5.265 billion in late November, 1998. It did so in spite of the fact Iraq had provoked another major crisis over UNSCOM inspections. Once again, the $5.265 billion figure will be more oil than Iraq can hope to export. Iraq's Deputy Oil Minister Taha Hamood stated on December 1, 1998 that Iraq would be unable to sell $3 billion worth of oil under a new six month phase of its oil-for-food deal. "The falling oil prices will reduce our oil exports even below the $3 billion worth of oil we had exported in the fourth phase of the deal. We are still having the same capacity of oil export. " He also said *that* Iraq was holding discussions with Syria on how *to reopen* an oil pipeline between them closed since 1982. The pipeline would have an initial capacity of 300,000 barrels per day and Iraqi oil officials say that $80 million is needed to repair the oil pipeline.

These figures do not include illegal oil sales through Iran and Turkey. Iran claimed in late 1998 to have largely halted the barging of product for transhipment through Iranian waters, although it is unclear whether this is true. The trade in diesel fuel between Iraq and Turkey was tolerated because it aided the Kurds inside their security enclave in Northern Iraq. It moved diesel fuel from Iraq's oilfields around Kirkuk through the Kurdish-controlled enclave to markets in Turkey. A Turkish official said in late November, 1998, that a total of 625 trucks that could carry up 10,000 liters of diesel each, had applied to Turkey to cross into Iraq that month.

The UN granted Iraq the right to import $600 million worth of equipment to begin renewing its oil production and export capacity. Progress, however, was slow and $600 million scarcely solves Iraq's problems. *By the middle of June, 1999, out of the allotted $600 million for oil industry supplies, the Sanctions Committee had only approved $273 million worth of equipment (mainly for the upstream) and put on hold contracts worth $74 million (mostly for downstream operations). However, only around $20 million worth of oil industry supplies had arrived in Iraq.*

On December 2, 1998, former Iraqi oil minister Issam al-Chalabi stated that Iraq required at least $5 billion to restore its crude oil production capacity to levels the existed before its invasion of Kuwait. He said that reservoirs, wells, storage facilities and pipelines all require urgent attention because of the impact of sanctions, and calls the $300 million approved for spares under the oil-for-food agreements "a drop in the ocean. " He said that Iraq also needed to invest in downstream facilities, and would require $30 to $50 billion to implement all of the plans it had developed before the invasion, including increasing production to 5.5 million barrels per day.

Iraq has already anticipated the full end to UN sanctions by signing oil and gas deals with companies from Russia, France, and China. In March 1997, for instance, a consortium headed by Russia's Lukoil signed a 23-year, $3.8 billion oil production-sharing deal with Iraq for development of the West Quma oilfield. French oil companies Total and Elf Aquitaine have negotiated all but the final details of development and production-sharing agreements for the Nahr Umar and Majnoon fields. Nahr Umar is estimated to have a potential production capacity of 440,000 barrels per day (*bpd*), while *Majnoon* contains estimated reserves of 7 billion barrels of crude oil. Upping the ante even further, Iraq in mid-October 1997 invited international partners to invest in natural gas projects worth $4.2 billion.

Russia is particularly interested in Iraqi oil deals. Russian Fuel and Energy Minister Sergei Generalov said on December 7, 1998 that, "Lifting of the present sanctions against Iraq would have both positive and negative implications for Russia, but positive effects would prevail. Russian companies would go on with their commitments to develop Iraqi oil fields, and Iraq would be able to service its debt to Russia. That would offset the biggest negative effect — more oil coming to international markets. " Generalov spoke during a visit by Iraqi Deputy Prime Minister Tariq Aziz, and said that Russian oil producers such as OAO Lukoil Holding, the country's biggest firm, and AO Tyumen Oil Co., the fifth-*largest*, were in talks with the Iraqi government to begin oil production in Iraq. Tyumen has bought Iraqi oil under a UN-sponsored oil-for-food program. OAO Gazprom, Russia's biggest company and natural gas monopoly, is in talks with Syria to rebuild a long-closed pipeline connecting Iraqi oil fields to a Syrian port and diverting Iraqi oil shipments from the only current route, which runs through Turkey.

It is unclear when sanctions will be lifted, or how much they will be eased in the future. There is still at least a moderate risk of a major military confrontation over the UNSCOM inspection dispute, and the U.S. has little incentive to agree to the lifting of sanctions in the foreseeable future. It is also unclear that the U.S. has any commercial interests that outweigh its security interests now *that* the oil for food deal is in place. Iraq is *not* likely to give U.S. firms high priority in the short-term unless if feels it can use them as leverage against U.S. policy. The mid- to long-term situation is less clear. Iraq has often badly miscalculated its own interests, but it normally is ruthlessly pragmatic and opportunistic within the terms of its own worldview. The current tensions with the U.S. will not preclude U.S. entry to the Iraqi market if Iraq feels that the U.S. can offer exceptionally favorable terms or has unique capabilities.

Sanctions on Libya

Libya shows as few signs of political moderation as Iraq, but it poses a much more limited threat and UN sanctions are punitive, rather than designed to enforce a cease-fire. At total of 270 people were killed in the 1988 bombing of Pan Am flight 103 over Lockerbie, Scotland. On April 15, 1992, the United Nations imposed economic sanctions on Libya for refusing to extradite two Libyan nationals accused of carrying out the attack. These sanctions included the grounding of all air traffic to and from Libya, a reduction in diplomatic relations and a ban on all arms sales to the country. In November 1993, UN sanctions were extended to include a freeze on Libyan funds overseas, a ban on the sale of oil equipment for oil and gas export terminals and refineries, and tougher restrictions on civil aviation and the supply of arms. These sanctions must be renewed at four-month intervals.

The United States later created its own sanctions against Libya, and attempted to persuade the UN to impose a total embargo on Libyan oil. On August 5, 1996, the United States imposed additional sanctions on Libya. This action—the U.S. Sanctions Act of 1996 — extended U.S. sanctions on Libya to cover foreign companies that make new investments of $40 million or more over a 12-month period in Libya's oil or gas sectors. However, several key European allies that relied heavily on Libya's low sulfur oil — including Italy and Germany — expressed deep reservations about this U.S. *policy*. Libya reacted by offering to have the suspects tried by the International Court of Justice or in a neutral third country, but has refused to hand them over for trial in the United States or Scotland. In late September 1997, the Arab League and the Non-Aligned nations called for an easing of sanctions on Libya. The United States responded that such efforts would not succeed unless Libya turns over the bombing suspects. In mid-October 1997, Libya officially asked the United States to hand over for trial U.S. pilots and officials involved in the 1986 bombing raid ordered by President Reagan. The U.S. State Department dismissed Libya's summons as "ridiculous." On March 20, 1998, the UN Security Council debated (without taking a vote) Libya's offer to have the suspects tried in a neutral court in exchange for the lifting of sanctions. It unanimously renewed sanctions on October 28, 1998.

Libya then made offers to allow Scottish judges to try the case in the Netherlands, and a long series of negotiations took place to determine if the offer was real and practical. Libya also circulated a report to the UN in early December, 1998, that claimed it had lost some $26 billion up to that time because of sanctions. Tahar Jehimi, the Governor of Libya's Central Bank warned the Libyan General People's Congress on December 8, 1998, that oil revenues would be 36% lower than in 1997, and would amount to only $5.8 billion. Libyan exports totaled $15.5 billion in 1990 in constant dollars, and averaged around $8 billion to $9 billion during the mid-1990s. As a result, Libya has had to make significant cuts in a number of entitlements and welfare payments and it suddenly devalued the Libyan Dinar by 18% on December 9. (From 0.38 Dinars to the dollar to 0.45.)

President Qadhafi released the two suspects on April 5, 1999 after months of negotiations and brokering by Saudi Arabian Ambassador Prince Bandar bin Sultan, South African President Nelson Mandela, Egyptian President Hosni Mubarak, and UN Secretary-General Kofi Annan. They now stand trial at a former air base in the Netherlands under Scottish law. In return for releasing the suspects, UN sanctions on Libya were suspended for ninety days, after which Secretary-General Annan reported to the Security Council on Libya's cooperation. If Libya complied with other conditions, sanctions could be lifted altogether. Although it agreed to the suspension of UN sanctions, the United States has declared that it will not lift U.S. sanctions or remove Libya from its official list of terrorist states until Libya complies with all the provisions of UN resolutions on Libya. The United States requested face-to-face talks with Libya so that they could decided whether to support the permanent lifting of sanctions.

Secretary General Annan reported his findings on Libya to the Security Council on July 2, 1999. According to Annan's report, Libya has, so far, complied with the provisions set forth by the Security Council. France has been satisfied with Libya's cooperation in regards to UTA Flight 772, and has convicted 6 people in absentia for the crime. Libya has provided assurances that it will continue to comply with the remaining requirements, including the compensation of victims' families in the event of a guilty verdict in the Pan Am 103 trial. The United Kingdom has stated that it has been satisfied with Libya's cooperation in combating terrorism, and hopes that Libya will continue to renounce terrorism. The United States, on the other hand, has not been as pleased with Libya's progress, and it does not seem likely that UN sanctions will be completely lifted until the end of the Pan Am 103 trial.

Libya presents a significant policy dilemma for the U.S.. On the one hand, it does not represent the kind of serious threat posed by Iran and Iraq, and sanctions are centered around acts of terrorism that could be relatively easy to resolve *since* Libya *has turned* over the two suspected terrorists. This makes the lifting of sanctions less complex than in the case of Iran and Iraq. It also would allow the rapid development of Libyan energy resources and a potential role for U.S. industry. On the other hand, the Libyan regime remains erratic and extremist, sanctions have had a major impact on Libyan efforts to proliferate, and they have also had a major impact on Libyan arms imports. Libyan arms deliveries have shrunk from an average level of about \$1.3 billion a year in the late 1980s to under \$10 million a year since 1993. Libyan new arms agreements dropped from \$3.5 billion in 1990-1993, to \$200 million in 1994-1997.

References

Soares de Oliveira, Ricardo. 2007. *Oil and Politics in the Gulf of Guinea*. New York: Columbia University Press.

Sprout, Harold and Margaret Sprout. 1971. *Towards a Politics of Planet Earth*. New York: Van Nostrand Reinhold

Turner, Louis. 1978. *Oil Companies in the International System*. London: Royal Institute of International Affairs

Watts, Michael. 2009. 'Crude Politics: Life and Death on the Nigerian Oil Fields'. *Niger Delta Economies of Violence Working Papers*. Institute of International Studies, University of California.

Wenger, Andreas; Robert W. Orttung, Jeronim Perovic (eds). 2009. *Energy and the Transformation of International Relations*. Oxford: Oxford University Press.

9

Changing Energy Policies

Middle Eastern states react to the impact of geopolitics and geoeconomics in very different ways. Each Middle Eastern energy exporting state must deal with the world market and respond to world oil and gas prices. Market forces, however, are only part of the pressures that shape national behavior, and key nations like Iran, Iraq, and Saudi Arabia face very different pressures even when they share common problems. Similarly, the smaller exporters also pursue individual national paths to energy development and to dealing with external threats and internal economic and political pressures.

Energy Policy and Development in Iran

Iran's energy policy, like its economic policy, is evolving towards greater reliance on market forces. At the same time, there are serious internal divisions with Iran's religious regime, and the theory and diagnostics of Iranian economic and energy policy are far better than the practice and actual solutions. Aside from the sanctions issue, major questions remain about Iran's political stability, its current efforts to attract foreign investment, and the quality of the National Iranian Oil Company's planning. These issues include:

- The constitutionality of any foreign participation in Iranian oil and gas fields and production.
- The quality of planning going into oil and gas development at the national, field, and venture level, which often seems to involve exaggerated hopes, non-economic scale, and poor analysis of cost and probable return.

- The realism of efforts to attract foreign investment, to guarantee an effective return on investment and repatriation of capital, and to limit investment risk.
- Contracting procedures which often begin with unrealistic demands, and which create uncertainties regarding Iran's reliability over time.
- A gas policy which fluctuates between efforts to export and the failure to establish realistic domestic prices and limit domestic demand.
- Major investments in nuclear power that consume scarce foreign capital, and which may well prove non-economic compared to gas-fired power plants.
- The overall distortion of energy *revenues* by the failure to move toward pricing energy at market prices.
- The uncertain short and medium-term balance of power between Iran's "moderates" under President Khatami and "traditionalists" under Rafsanjani, and the resulting energy and investment policies Iran will follow.
- Iranian willingness to re-price the domestic cost of gas and oil to limit the growth of domestic demand and ensure that sufficient gas and oil will be available for export.
- Iran's ability to carry out the broader economic reforms necessary to ensure even its short term political and economic stability, and ability to meet its debt payments.

At the same time, there is no question about the importance of Iran's energy reserves and energy production. Iranian oil and gas will play a major role in shaping future world energy supplies.

Iranian Energy Export Income and the Economy

Iran has a more diversified economy than most Middle Eastern oil exporting states. Industry accounts for 37% of its GNP, and services for 42%, *while* only 21% of its GNP is devoted to agriculture. It has major industries for petrochemicals, textiles, cement and other construction materials, food processing, metal fabricating, and armaments. As a result, Iran needs a diversified and sophisticated range of imports. Iran also has a major backlog of potential imports because of the prolonged impact of the Iran-Iraq War (1980-1988) and revolution (1979 to present) on Iran's ability to finance the imports it needs.

Iran's economy is currently operating under a five-year plan, which ends in 2000. The economy grew strongly in 1997, partly due to relatively high oil prices during the first two-thirds of the year The Department of Energy estimates that oil exports reached $15.7 billion and accounted for 80%-85% of total export earnings in 1997. Iran's real gross domestic product (GDP) grew 3.4% in 1997. Even so, Iran's GNP was almost the same value in 1997, measured in constant dollars, as it had been in 1985. Combined with massive population growth, Iran's economic problems caused its per capita income to drop from over $3,500 in 1995, to under $2,500 in 1997 ($5,300 in purchasing power parity terms). Iran was hit hard by the "oil crash" with oil exports dropping to $10.2 billion, and the GDP growth rate falling to 2.9%.

Iran's economy had experienced years of austerity before the economic growth of 1997, however, because of the impact of the Iran-Iraq War and sanctions, the stringent import controls necessary to allow Iran to repay foreign debt, the corruption and inefficiency of the government and large religious foundations (Bunyods), and the failure of agricultural reform and growing dependence on food imports. Its growth in 1997 was the exception, not the rule, and it occurred as much because Iran's economy had been weak in the past as because of high oil revenues *during* much of 1997.

Low oil prices and revenues made Iran's economic situation much worse in 1998. Iran originally shaped its budget for the fiscal year beginning March 21, 1998 on estimated oil prices of $17.50 per barrel. This was revised downwards in January 1998 to $16 per barrel, and then again to $12 per barrel effective on March 21, 1998 (the beginning of *the* country's new fiscal year). Iran based its budget for the fiscal year beginning March 21, 1999 on an estimated $11.80 per barrel. Iran reported that its earnings from oil and gas exports fell by 20% in the year between March 20, 1997 and March 21, 1998, although this period includes nearly six months before the drastic fall in oil prices. As a result, oil fell from 57.2% to 41.6% as a share of state revenues. This forced Iran to cut its imports by 2.6%. For 1998 as a whole, Iran's oil revenues were estimated to be about $10.2 billion, down 35% from $15.7 billion in 1997. Iran's oil revenues for 1999 are expected to rise to $12.1 billion, still 23% lower than pre "oil crash" levels.Iran's economy has been crippled by factors that have long-term implications, regardless of which side wins the current struggle for political power and the day-to-day price of oil.

Economic Reform with Uncertain Plans, Action, and Money

In March 1998, President Khatami said that "the structure of our economy is sick " and that "we have to get used to spending less hard currency. " Iran's government is seeking to keep inflation under control, largely through a tight fiscal policy. Repayment of Iran's large foreign debt also remains a priority (and a drag on economic growth), with debt service in 1998 made more difficult by projected lower-then-expected oil prices (oil export revenues account for around 40% of government revenues). In January 1999, Iran's Majlis (parliament) set its oil price forecast for the fiscal year beginning March 21, 1999 at $11.80 per barrel. Iran claims it loses $1 billion for each $1 per barrel drop in the price of its oil. As a result, Iran will probable experience negative real growth in 1999.

The "oil crash" of 1997-1998, and continuing low oil revenues in 1999 have interacted with Iran's political crisis to virtually paralyze substantive moves towards economic reform. Iran has, however, taken some steps toward reform. In December 1997, the Iranian government announced that it would raise gasoline prices at the pump by 25% in the next fiscal year (beginning March 21, 1998). Subsidies on oil products amounted to some $11 billion a year in 1997 (twice the country's development budget), placing significant strains on the national budget, contributing to widespread inefficiencies in the Iranian economy, and increasing domestic demand for oil (thus reducing the amount left over for export). Iran acted in part because the International Monetary Fund (IMF) was pushing Iran to cut subsidies *in order* to liberalize trade and taxation, and to unify the country's two-tier exchange rate system. Iran has also made several price increases. In March 1999, gasoline prices increased by more than 70%, and fruit and vegetable prices increased by more than 80%. In mid-March 1998, the Iranian government announced plans to privatize about 2,400 state-owned companies, including several in the oil, petrochemical, and other energy sectors, though real progress has been slow.

President Khatami was elected largely on a platform of social justice, however, and has increasingly had to balance efforts to reform the economy against political pressures to keep unemployment low and to avoid social unrest. The resulting lack of progress may also have been one of the factors causing student riots in June 1999, and many Iranians believe the economy will seriously undermine popular support for Khatami unless things improve within the next year.

Iran recognizes that foreign investment and technology are important potential solutions to its problems. President Khatami has given speeches stating that he seeks to make Iran more attractive to foreign investment, particularly in the critical oil and gas sectors. Iran's top priorities for improving the investment climate are increased protection from nationalization and easier repatriation of profits. The Khatami administration predicted $7 billion of new foreign investment in the oil, gas, and petrochemical sectors during the new fiscal year beginning March 21, 1998, and up to $94 billion total through 2010.

Khatami and his advisors realize, however, that there is virtually no chance that such investment will take place without a massive restructuring of the Iranian economy. President Khatami set forth his proposals for such a restructuring on August 2, 1998. He based his plan on the recommendations of a commission of experts appointed in September, 1997, that included key ministers, the Governor of the Central Bank, and Head of the Plan and Budget Organization. The commission evidently split sharply, *which* delayed action for several months. When Khatami spoke, however, he did so after winning the support of Khamenei.

While Khatami's speech stressed the need for social justice, he again warned that, "the economy is sick and should be treated at the roots. " He said that the drop in oil revenues was pushing the need for reform towards the crisis level, and that a successful plan had to be implemented as an integrated effort, and not piecemeal.

Khatami's speech provided more diagnostics and good intentions than practical plan. Although Khatami promised the Plan and Budget Organization would soon provide full details, these details are still unavailable. So are any meaningful signs of effort from the task force Khatami set up to implement his reforms. The Iranian Majlis, for example, has not even been able to begin debate on price reforms in the energy sector that *would* help force Iran to import some $500 million a year in petrol. Iran would have to pay 1,000 rials a liter for the imported petrol, even though petrol sells for only 200 rials a liter in Iran, and Iranian petrol smuggled into Afghanistan sells for 1,750 rials a liter.

It is also unclear how much time and flexibility Iran really has in carrying out its proposed reforms. Its budget deficit rose to $6 billion in 1998. At current oil export levels, Iran cannot pay for its debt and key entitlement problems, much less reform. Inflation and unemployment are up and construction is down by about 40% compared to 1997. The trade balance has dropped from a $7.4 billion surplus in 1996-1997 to $3.05 billion in 1997-1998, and is expected to be $2.43 billion in 1998-1999. Iran is experiencing a steadily larger budget deficit, depreciation of the Iranian rial, and a shortage of foreign exchange without being able to predict the scale of the problem.

Debt Repayments and Fiscal Discipline

Iran faces immediate problems in terms of debt repayments and cash flow, largely as a result of the disastrous economic reform plan proposed by former President Rafsanjani and his technocrats after the end of the Iran-Iraq War. This led to massive foreign borrowing that did little more than fund massive corruption. The resulting payments problem triggered the need for strict controls on imports and hard currency that has crippled Iranian economic development for most of the early and mid-1980s.

Khatami and his ministers have emphasized the importance of fiscal discipline and of balancing the budget ever since Khatami's election in 1997. Iran paid $4.7 towards its debt obligations during 1997/1998. There is some uncertainty over the size of Iran's remaining foreign debts. The government stated that its foreign debt was $12.1 billion at the start of 1998, with $3.3 billion short term and $8.8 billion medium to long-term. Iran must repay $3.05 billion in 1998-1999, $864 million in 1999-2000, $481 million in 2000-2001, $256 million in 2001-2002, and $530 million in 2002-2003. In March 1998, however, Iran's Central Bank Governor, Dr. Mohsen Nourbakhsh, estimated that Iran had $26.4 billion in foreign debt obligations, including $14.1 billion in confirmed debts. Iran has since been forced to reschedule and seek at least $3 billion worth of refinancing. Iran is also negotiating barter contracts for food imports from nations like Australia and France that effectively mortgaged 1999's oil for 1998's wheat imports.

By October 1998, Iran had to ration the payments on the debt it had already rescheduled in 1994. It did so although it had already been forced to cut its budget by up to 20% in many areas. The basic problems were that Iran's hard currency income for 1998/1999 was now projected at around $14 billion, but it had an import bill of $15 billion, driven largely by entitlements, welfare, and state capital expenditures that were extremely hard to cut. It also had debt repayments of $4.0 -$5.0 billion. Iran was able to secure rescheduling of $2 billion of its foreign debt in 1998, but was still looking for another $1.5 billion in February 1999, and was seeking to reschedule over a period of 33 months, rather than 12. Moshen Nourbaksh, the Governor of Iran's Central Bank, put Iran's total debt at $13 billion in March 1999, but noted that it owed another $11 billion in purchases made through the Central Bank.

While Iran had enough foreign exchange reserves to get through most of the next year ($6.7 billion in March 1998), it regarded these as strategic, and it faced a forecast budget deficit of $6.3 billion. The sums were scarcely critical for a nation the size of Iran, but Iran's Majlis opposed any draw down of its "core reserves" and forced Iran to go to Germany and Japan for rescheduling.

Unless some unforeseen factor leads to major increases in oil prices, Iran will face a serious and continuing economic crisis. Iran is, however, trying to deal with some of its structural economic problems. The Majlis approved a bill in January 1999 to raise domestic fuel prices by up to 40% in current Iranian rials. It is a symbolic of the political problems of reform, however, that the Majlis agreed to a 75% rise in premium gasoline beginning in March, 1999, but would not allow any rise in the price of regular gasoline.

Maintaining and Expanding Iranian Oil Production

The decline in Iran's oil export revenues has had an obvious impact on much-needed investment in the country's oil sector. As a result, Iran's moderates and pragmatists have looked towards western capital markets as a source of investment capital. For example, President Khatami and oil minister Namdar Zangeneh have decided to open previously forbidden onshore oil fields to foreign oil companies, which have been barred by the constitution from the strategic oil sector since the 1979 Iranian revolution. The Expediency Council has also approved a law protecting investments from nationalization and arbitrary administrative decisions.

The practical impact of such policies is uncertain. As has been discussed earlier, foreign companies increasingly regard U.S. third party sanctions as hollow Congressional posturing that the Clinton Administration will be *quick* to waive. *This perception* led to widespread enthusiasm when Iran tendered 47 oil and gas projects in July 1998. By December, however, many companies had begun to question whether the words of Iran's moderates were cost-effective in terms of practice. This was partly the function of low oil prices, but most Western companies wanted direct equity so they could book the resulting oil reserves as assets and boost their share prices.

The constitution Iran adopted in 1979 prohibits foreign concessions, and only permits "buy back" contracts in which the investing company advances capital and is repaid with a fixed profit margin from the sale of oil and gas produced by the project. Such deals have the advantage that Iran must bear all of the risk relating to oil prices, but they effectively treat the foreign company as a service company with no long-term role in a given oil or gas field. It is not clear that such terms are competitive enough to meet Iranian needs – particularly given the fact that any legislation protecting foreign companies today could be overruled by a future change in power.

Iran has made the development of such buy back deals a high priority, and they have received considerable support from the Iranian Majlis. In spite of Iran's growing economic problems, the Majlis approved the funding of $5.4 billion in buy back projects for the 1999/2000 fiscal year, the same amount approved for 1998/2000. The NIOC has also requested that no ceiling be put on gas projects, and it seems likely that the Majlis will approve this.

The NIOC offered 26 development projects for bid in July 1998, and 17 exploration blocks. Iran then steadily increased the number of projects it has tendered from 42 in July 1998 to 47 by the end of the year. Iran is now seeking to make early awards for 30 of the 47 projects it has developed for tender, some of which are important if Iran is to preserve production by current fields and maintain overall production by developing new ones. It has also has tried to expand the range of countries under contract.

The National Iranian Oil Company (NIOC) awarded a $19 million contract to Royal Dutch Shell and LAMSO (UK) in late 1998 to explore for oil in the Caspian. This was the first upstream contract awarded since 1979, and British Petroleum was given the option of joining the consortium later – a way of deferring the sanctions issue. Its high priorities include development of Darkhovin, near the Iraqi border, development of phases 4 and 5 at the South Pars field, improving recovery at the Agha Jari field where output has dropped to 200,000 b/d, and gas reinjection at the Bangestan resevoirs in the Ahwaz area. It is also finalizing downstream projects. A 390-kilometer pipeline from Neka in the Caspian to Tehran has been awarded to the Iranian firm of Mapna, and a 35,000 b/d condensates project at the Bandar Abbas refinery has been offered for bids by Asian and local firms. In late 1998, however, Iran only managed to finalize three of the 11 projects it had first tendered in 1995. A $600 million investment in the offshore Doroud oil field, a $150+ million investment in the offshore Balal oilfield, and the onshore NGL1200 and NGL1300 gas projects. Once again, it diversified its suppliers. The Doroud project was won by Elf Aquitaine of France and Agip of Italy, the Balal project by Bow Valley Energy of Canada and Premier Oil of the UK, although the NGL project went to the Iranian Oil Industry Engineering Company. However, in March 1999, Premier failed to sign the Balal oil field service contract, and the NIOC awarded its share to Elf Aquitaine.

Developing Iranian Oil Reserves and Raising Production

Iran has large reserves and should remain a major potential oil producer well beyond 2020. The EIA estimates that Iran holds 93 billion barrels of proven oil reserves, or roughly 9% of the world total. The vast majority of these proven reserves are located in giant onshore fields in the Khuzestan region near the Iraqi border and Persian Gulf terminus, and more than half of Iran's 40 producing fields contain over one billion barrels of oil. Iran has not had significant oil exploration activity in nearly 3 decades and its actual reserves could be substantially larger.

Most of Iran's crude oil is low in sulfur and light, with gravities in the 30°-39° API range. Its current sustainable production capacity is estimated at around 4 MMBD, although this figure is controversial and some experts claim that Iran has only maintained production levels at several older fields by using methods which have permanently damaged the fields. In any case, Iran is OPEC's second largest oil producer. Iran produced an average of 3.1 million barrels a day in 1990, 3.3 million in 1991, 3.4 million in 1992, 3.5 million in 1993, 3.6 million in 1994, 3.6 million in 1995, 3.7 million in 1996, 3.7 million in 1997, and 3.6 million in 1998.

The onshore Ahwaz, Marun, Gachsaran, Agha Jari, and Bibi Hakimeh fields alone account for about two-thirds of Iran's oil production. Iran produces about 500,000 barrels per day of crude oil from eight operational offshore fields. This output level is the result of extensive development and refurbishment efforts by the National Iranian Oil Company (NIOC), which have boosted offshore capacity by about 150,000-200,000 barrels per day since 1993. The Doroud 1&2 (146,000 barrels per day), Salman, Abuzar, Forozan, and Sirri C&D fields comprise the bulk of Iran's offshore output, all of which is exported. Iran plans extensive development of existing offshore fields and hopes to raise its offshore production capacity to 1 MMBD by 2000.

Iranian officials estimate that the country will need to invest as much as $90 billion in its oil industry over the next decade *if* it is to avert a dramatic drop in oil production, although many estimates go as low as $40 billion. *It is estimated that* development of new offshore Persian Gulf and Caspian Sea oil fields will require investment of $8-$10 billion. In December 1997, Oil Minister Zanganeh stated that the country aimed to boost oil production capacity 200,000-250,000 barrels per day each year. He set a goal of reach 5 MMBD in the next five years, possibly surpassing 6 MMBD by 2010, and reaching 7 MMBD at some point between 2010 and 2020. These goals would more than restore the country's production capacity to the level of over 6 MMBD it achieved in the mid-1970s. Iran estimated that meeting these goals would require anywhere from $8 billion to $30 billion worth of investment.

U.S. government experts are only slightly less optimistic. Many experts question whether even six million barrels per day is possible, however, and the majority of industry opinion seems to be that four to five million barrels a day may represent the limit of sustained production of conventional oil production even if Iran can fully fund its goals for domestic and foreign investment.

Since President Khatami's election, Iran has attempted to coordinate oil policy with Saudi Arabia. Iran supported Saudi Arabia in leading OPEC to decide in November 1997 to raise its production ceiling by 10%. The end result was a major fall in oil prices due to a warmer-than-normal winter in the northern Hemisphere, lower Asian oil demand, and OPEC overproduction. As a result, both countries have had to make major production cuts and face serious economic problems. This led to Saudi changes in October and November, 1998, that Iran was systematically violating its production quotas. The two countries seem to have reached a new agreement on lower production goals in late November.

During OPEC meetings in March 1999, Saudi Arabia focused heavily on resolving the dispute over Iran's compliance. An agreement was finally reached, delimiting the baseline for Iran's share of OPEC cutbacks at 3.9 million barrels per day, rather than 3.6 million barrels per day as argued by other OPEC member states. Although Iran will face the same 7.3% reduction as other member states, in actuality Iran's cuts will be smaller due to the amended baseline.

Iran is attempting to modernize its oil industry and discover new fields. In January 1999, Khatami approved a plan to restructure the industry, including decentralization and the separation of policy from executive affairs. According to press reports, the plan calls for "fundamental changes in management." In addition, the NIOC has focused on frontier exploration efforts in the hopes of adding 1-2 billion barrels of proven reserves, which would also allow Iran to increase its daily oil production.

There already are some indications that Iran may seek even more ambitious long-term production goals as a result of the current crisis in oil prices and changes in Saudi strategy. On November 18, 1998, Hossein Kazempour Ardebili, a senior adviser to the oil ministry, stated that Iran would have to boost its crude production capacity if it hopes to attract foreign investment in its oil and gas industry:

"In the future oil producing countries will possibly be forced to ease off conditions for foreign investment, therefore Iran must do its utmost to attract investments by increasing its production capacity. In the years to come, Saudi Arabia's production capacity will reach 15 million barrels per day that will be alarming for Iran as regards security in oil markets. Boosting production capacity would strengthen Iran's bargaining position against other oil producing countries.In the interim, however, Iran is bound by OPEC production quotas, and is currently allowed to produce only 3.359 MMBD.

Iranian Crude Oil Exports

Iran exported about 2.4 MMBD in 1997, mainly Iranian Heavy (31° API) and Iranian Light (34° API) blends. Exports of Iranian Light slowed in 1997, however, and Iran attempted to shift term customers onto Iranian Heavy. Smaller blended crude streams come from the Lavan, Sirri, and Forozan area fields. *Since* early 1998, Iran's core Asian customers (Japan and South Korea) have purchased less oil than normal from Iran, mostly due to the Asian economic crisis. In response, Iran has increased its marketing efforts in Europe, Latin America, South Africa, and China.

Iran has not been able to modernize its export facilities because of war and cash flow problems. All Iranian onshore crude oil production and output from the Forozan field (which is blended with crude streams from the Abuzar and Dorood fields) is exported from the Kharg Island terminal located in the northern Gulf. The terminal's original capacity of 7 MMBD was nearly eliminated by more than 9,000 bombing raids during the Iran-Iraq War. Kharg Island's current export capacity is 5 MMBD, or double Iran's 1996 total crude oil exports of 2.5 MMBD.

Iran installed four single buoy moorings at Ganeveh during the Iran-Iraq War. These provide an additional combined capacity of 1 MMBD. Furthermore, smaller amounts of offshore crude oil production from the southern Persian Gulf are exported from terminals on Lavan Island and Sirri Island. Iran also has unused terminals at Cyrus and Ras Bahregan in the southern Gulf. In June 1995, the 60,000-dwt capacity Abadan product terminal was re-opened after having been shut-*down* since the Iran-Iraq War.

Like Saudi Arabia, Iran has sought to create strategic oil reserves, but it has had substantially less success. The NITC has acquired five 1,300,000-dwt tankers from South Korea's Daewoo. These tankers will complement NITC's existing fleet of six very large crude carriers (VLCCs) and will allow NITC flexibility in marketing Iranian crude. The remaining five vessels continue to be used as floating storage vessels for offshore Iranian loading terminals in the Persian Gulf. In June 1999, NITC awarded a $410 million contract to Hyundai Heavy Industries of South Korea for 5 VLCCs.

In July 1995, South Africa agreed to allow Iran to store up to 15 million barrels at the 45-million barrel Saldanha Bay crude oil storage facility near Cape Town. In January 1996, however, South Africa's Central Energy Fund reported that an environmental study designed to assess the impact of storing Iranian oil in South Africa would be delayed by six months until September 1996. In late 1996, the South African government announced that extensive environmental studies would have to be undertaken before the storage deal could move forward. Since no firm timetables have been established for the completion of these studies and Iran now faces major cash flow problems, the Saldanha Bay storage deal is likely to be delayed indefinitely. The deal would have enabled Iran to store crude exports closer to its South American and European markets. It also would have served to facilitate supplies to two South African refineries in Durban, which lifted about 200,000 barrels per day of Iranian crude in 1995.

Iranian Product Imports

Iran has made important progress in one aspect of its energy problems. The

Iran-Iraq War destroyed much of its refinery capacity, and left it dependent on product imports, including gasoline, kerosene, and gasoil. As a result, Iran imported about 131,000 to 165,000 barrels of product a day during 1993-1997, even though it also exported 105,000-136,000 barrels a day of fuel oil. The completion of a new refinery at Bandar Abbas gave Iran 230,000 b/d of additional product output in 1997, and the capacity of the refinery at Abadan was raised by 50,000 b/d. This raised Iranian capacity to 1.542 MMBD. Actual output has risen from 1.089 MMBD in 1993 to around 1.45 MMBD. As a result, Iranian imports dropped to around 31,400 b/d in 1998.

The Caspian Sea

Iran is only one element in the complex mix of issues affecting the Caspian Sea and Central Asia, but its role is significant. The Department of Energy indicates that its estimates of the Caspian Sea's proven and possible oil reserves could reach 191 billion barrels, along with huge natural gas reserves. Since the break-up of the former Soviet Union, territorial issues have arisen regarding rights to the Caspian's resources. Iran's position is that treaties signed in 1921 and 1940 are still valid, implying that all countries bordering the Caspian must approve any offshore oil developments.

In late February 1998, Iran's Foreign Minister Kamal Kharrazi reiterated Iran's position that any unilateral exploitation of Caspian Sea resources would be illegal. Oil Minister Zanganeh has stated that Iran backs national zones extending several miles from the coast and a "condominium" in the middle of the Sea. Iran also has stated (along with Russia) that it opposes laying an oil pipeline across the Caspian Sea floor.

The pipeline issue is a major source of controversy between the U.S. and Iran. Iran feels it is the natural transit route for oil and gas exports from the Central Asian countries to world markets. U.S. policy opposes pipelines through Iran, the shortest (and most likely the least expensive) path to the open sea. The United States has instead favored multiple routes for Caspian oil and gas through the Caucasus region to the Black Sea or to the Turkish port of Ceyhan, as part of its attempt to isolate Iran and to contain its influence in the region.

Several U.S.-based oil companies feel this policy is wrong and that U.S. sanctions are forcing American companies to opt for less preferable and accessible ways of moving oil and gas from the Caspian region to market.

Unocal, for instance, has proposed building expensive, large-scale oil and natural gas pipelines from Turkmenistan to Pakistan through war-torn Afghanistan, instead of through Iran, because of sanctions, even though it believes a route through Iran would make more sense economically. Other companies feel that Iran is the natural commercial route for pipelines to the Gulf. The French firm, Total, sees Iran as the logical pipeline route to carry oil from its projects in Azerbaijan. BP Amoco, which heads the main consortium developing oil reserves in the Caspian Sea off Azerbaijan contends that there will have to be a major export route through Iran, otherwise oil companies in the region would "have a problem."

The NIOC is currently exploring Iran's territorial waters in the Caspian Sea and has sought stakes in several of the various oil field development projects offshore Azerbaijan. Since 1995, Iran has been conducting a five-year exploration program of its sector of the Caspian Sea. However, exploration largely has been limited to shallow waters and primarily has resulted in marginal natural gas finds. In late 1996, National Iranian Drilling Company and Azerbaijan's SOCAR formed a joint venture company to conduct seismic surveys and drill exploratory wells in the Iranian sector of the Caspian. In February 1996, Turkmenistan invited Russia and Iran to conduct exploratory drilling in that country's Caspian Sea sector, where oil reserves are estimated at around 100 million barrels. In December 1998, Royal Dutch/Shell and Lasmo reached a deal with NIOC to conduct seismic exploration of Iran's share of the Caspian Sea. The $19-million project will allow the consortium to negotiate service contracts after the geological study is completed. Azerbaijan immediately protested the decision, claiming that some of the area to be studied is within its territorial waters. Iran disputed the charge, but Azerbaijan raised the stakes by opening tenders for three offshore fields in the disputed area. In response, NIOC established the Khazar Exploration and Production Company (KEPCO) to manage oil and gas operations in Iran's Caspian *Sea sector.*

Crude Swaps

Iran has pushed so-called "swap" arrangements as part of its strategy to promote itself as an export route for oil and gas production from Kazakhstan and other Central Asian states. The Department of Energy reports that Iran and Kazakhstan agreed to such an arrangement for Kazakh crude exports in May 1996. The swap deal is for 10 years and involves about 70,000 barrels per day. Initial swaps began in December 1996 and involved shipments across the Caspian Sea of crude oil from the Tengiz, Kalamkas, and Karazhanbas fields, blended to produce a 32.5° API gravity. This blend is comparable to the Iranian crude feedstock currently utilized at the 225,000-barrels per day Tehran and 112,000-barrels per day Tabriz refineries in northern Iran. In exchange for Kazakh crude supplied to northern Iranian refineries, Kazakhstan will have the option of lifting either Iranian Light or Iranian Heavy blend crude at Kharg Island. The swap agreement was temporarily put on hold in December 1997 due to high sulfur content and other " technical problems" with Kazakh crude.

Much larger swap arrangements may occur in the future. Iran claims that it could handle 750,000 barrels per day in Caspian crude swaps in a short period of time, and up to 1.5 MMBD over time. According to Iran, this would involve "not much investment"—mainly in modifying existing pipelines linking the Tabriz and Tehran refineries with the Caspian coastal cities of Astara, Bandar Anzali, Rasht, Neka, and Gorgan. Modifications would entail a partial reversal of one or more product lines, implementation of a means to alternate between crude shipments south from Neka and product shipments north, or construction of a new crude line from either the Neka or Bandar Anzali ports to the Tehran or Tabriz refineries, respectively. (In early March 1998, plans were announced to expand a pipeline from Neka to the Tehran refinery.)

The U.S. government rejected a request by Mobil Oil Corporation to swap crude oil from the Caspian Sea region with Iran. Mobil applied for the permit in 1998, and has been one of the leaders in a campaign by U.S. companies to change U.S. sanction policy. The company holds a 40% stake in a Turkmenistan joint venture that produces 12,000 bpd of crude oil, most of which Mobil had hoped to move through Iran.

Swap arrangements make sense for Iranian domestic purposes, as well as creating the equivalent of Caspian oil exports through Iran. Most of Iran's oil is located in the south, far from major population centers and refineries in the north, meaning that large volumes of oil have to be pumped long distances across Iran. "Swaps" could help address this problem.

Natural Gas

Iran's natural gas development will be a key factor shaping its future energy policy and exports, and possibly its nuclear policy as well. The EIA estimates that Iran contains 812 trillion cubic feet (Tcf) of natural gas reserves—the world's second largest and surpassed only by those found in Russia. The bulk of Iranian gas reserves are located in non-associated fields. However, large onshore oil fields contain approximately 120 Tcf of associated gas, which is either dissolved in crude or is in gas caps. Iran's largest non-associated gas is the South Pars field, which is an extension of Qatar's 241-Tcf North Field.

South Pars was first identified in 1988 and was originally appraised at 128 Tcf in the early 1990s. NIOC-sponsored studies conducted in mid-1996, however, indicate that South Pars contains an estimated 240 Tcf, of which a large fraction will be recoverable, and at least 3 billion barrels of condensate. Iran's other sizable non-associated gas reserves include the offshore 47-Tcf North Pars gas field (a separate structure from South Pars), the onshore Nar-Kangan fields, the 13-Tcf Aghar and Dalan fields in Fars province, and the Sarkhoun and Mand fields.

In 1996, Iran produced about 2.6 Tcf of natural gas. Of this amount, 1.3 Tcf was marketed, 1 Tcf was re-injected, and 0.3 Tcf was flared. In 1990, Iran undertook an ongoing gas utilization program that is designed to boost production to 10 Tcf per year by 2010, reduce flaring, provide gas for EOR re-injection programs, and allow for increased gas exports abroad.

Although domestic gas consumption is growing rapidly, Iran continues to promote export markets for its natural gas. Under current plans, NIOC optimistically hopes to export 450 Mmcf/d of gas by 2000, rising to 4,000 Mmcf/d by 2005 as its larger, more ambitious projects come online. By 2000, NIOC plans to have completed three gas export pipelines to Turkey, Armenia, and Nakhichevan. By 2005, two more lines to Europe and India are planned, in addition to the possibility of a liquefied natural gas (LNG) facility for LNG exports to Asia. Implementation of these ambitious plans will require substantial international financing and support, both of which are lacking at present. Iran also hopes to serve as a major transit center for gas exports from Central Asia.

Natural Gas Exports

In August 1996, Iran signed a $20-billion, 22-year gas supply deal with Turkey. Initial gas deliveries of 300 Mmcf/d are expected to begin in 1999, with later supplies climbing to near 1 Bcf/d. The supply contract required the construction of four new pipelines. As of January 1997, design and engineering work had been completed for the first link between the Iranian Gas Trunkline (IGAT 1) terminus at Tabriz and Bazargan on the Turkish border. On the Turkish side, state-owned Botas recently began the bidding process for construction of a 200-mile, 40-inch link between Dogubeyazit and Erzurum. In April 1997, Botas requested bids for a 252-mile link between Erzurum and Sivas as well as a connecting 288-mile pipeline between Sivas and Ankara. Iranian gas used to supply the pipeline will come from the non-associated Kangan regional fields as well as from associated sources around Ahwaz. *In November 1998, Turkey began construction of a 623-mile pipeline that could transport gas westward from Iran. The Ezurum-Ankara line could also be used to transport gas from a pipeline from Turkmenistan running beneath the Caspian Sea. Iran has also proposed the construction of the IGAT 3 pipeline from South Pars to Turkey, which could cost $2-3 billion.*

This deal brought new criticism from the United States, which viewed it as a major source of economic support to Iran. Turkey made it clear, however, that it needed to diversify its suppliers of natural gas and that Iranian gas was one of its most economically sound alternatives. Since its deal with Iran, Turkey has signed a $3 billion dollar deal to buy large volumes of gas from Turkmenistan (possibly via Iran) beginning in 2000.

In mid-1995, Iran signed a renewable, 15-year deal to supply 100 Mmcf/d of gas to Armenia. This agreement required construction of a new $135-million pipeline link that connected to IGAT 1. Construction of the line is proceeding, and first gas exports are expected to occur by 1999. The Iranian gas will be used to supply a new, jointly constructed power plant on the Aras River.

In December 1997, Turkmenistan, Iran, and Turkey signed an agreement with Shell for a nine-month feasibility study into the possibility of constructing a natural gas pipeline that would eventually link Turkmenistan to Turkey (and beyond) via Iran. The $1.6-billion, 930-mile long pipeline initially would carry 105 Bcf of gas per year. In early March 1998, the first natural gas began flowing through a relatively small, $195-million, 125-mile pipeline from Turkmenistan's southern Korpedzhe gas field to Kurt-Kuyi, Iran. The line, inaugurated on December 29, 1997, was financed by NIOC (80%) and the Turkmen government (20%). Initial capacity is 141 billion cubic feet per year, expected to double by 2006. Both Iran and Turkmenistan hope that this small initial line will eventually be part of a much larger regional gas pipeline network. Turkmenistan initially will sell its gas to Iran at a price of $40 per thousand cubic meters ($1.13 per thousand cubic feet).

Iran is increasingly targeting *emerging* Asian markets like Pakistan and India (rather than Japan and South Korea) for LNG exports. In January 1995, Iran and Pakistan signed a preliminary agreement for construction of a $3-billion, 870-mile, 1.6-Bcf/d onshore gas export pipeline linking South Pars with Karachi, Pakistan. Subsequently, a proposed extension to India led to security-of-supply concerns for the Indian government. As of February 1998, a consortium of Shell, British Gas, Petronas, and an Iranian business group known as the "Foundation for the Deprived and the War Disabled " reportedly was negotiating to export gas from South Pars to Pakistan. A committee was established in April 1999, to examine the possibility of a gas pipeline from Iran to India via Pakistan.

In May 1999, British BG proposed construction of a one billion LNG facility on Kish Island to supply gas to India. The proposal calls for construction of LNG production, storage, and export facilities on the Island, located in the Central Gulf. The project is in the pre-feasibility stage, and awaits gas supply guarantees by the NIOC and a government assessment of environmental risk factors.

Nuclear Power

Iran's search for nuclear power affects several key aspects of its energy development, including sanctions against proliferation, domestic consumption of oil and gas, and surplus energy exports. To put the issue in terms of Iranian hopes and plans, Iran hopes to raise its electrical generating capacity from 25,000 megawatts (MW) at present to 30,000-MW by 2000. Nuclear power is only part of its effort to meet this goal. Hydroelectric generation capacity is projected to nearly double to 5,000 MW after construction of a dozen new dams during the next several years.

Iran relies on continuing foreign investment for a number of other ongoing power projects. These include a Kraftwerk Union (a Siemens' subsidiary) $1.4-billion, 2,080-MW combined-cycle plant in southern Iran. Other foreign companies active in Iranian power sector projects include Canada's Babcock & Wilcox, French-British GEC-Alsthom, Switzerland-based ABB Asea Brown Boveri, and Italy's Nuoca Cimimontubi and Belleli. In January 1998, Iran's Deputy Energy Minister said that Iran would welcome European and U.S. private investors to participate in the planned privatization of the country's power generation industry. Breaking up the state power generation monopoly (Tavir) into competing private companies and reducing large state subsidies are two important proposed measures aimed at increasing electric generation and transmission efficiency, reducing an estimated $4 billion in wasted electricity, and attracting foreign investment.

Nuclear power does, however, remain a major goal and has been one since the time of the Shah. Work began on a nuclear power reactor at Bushehr in 1974, but was halted (80% complete) following the 1979 Islamic Revolution. In January 1995, work on Bushehr resumed after Russia signed a $780-million contract to complete the power plant. The Russian-Iranian deal called for completion of the two 1,300-MW pressurized-light water units and an option for two modern VVER-440 units.

The United States strongly opposed the project, and provided Russia with intelligence information pointing to the existence of an Iranian nuclear weapons program. There have been numerous U.S. attempts to halt the Russian deal and deny Iran other sources of nuclear technology. Recent attempts have been made by both Secretary of State Madeleine Albright and Vice-President Gore. On March 6, 1998, for example, the Ukraine announced that it would not sell turbines for use with reactors at Bushehr during a visit by U.S. Secretary of State Albright,. The contract had been worth $45 million. Five days later, Vice President Gore met with Russian Prime Minister Chernomyrdin and discussed U.S. objections to Russian exports of nuclear and missile technology to Iran.

The most serious U.S. action, however, occurred in January 1999 when the U.S. imposed sanctions against two Russian nuclear research centers. These centers experienced serious financial difficulties, and in March 1999, Russia offered to cut back its nuclear aid to Iran if Washington ended its sanctions against the two Russian institutes.

Russia, however, continued work on Bushehr, and some estimate that the plant could be completed by 2000-2003. The Iranians renegotiated the structure of their contracts for the first reactor with Russia during the spring and summer of 1998, and it is now virtually a turnkey project. Russian's are now fully responsible for every major critical aspect of the program.

The first reactor may also be only the first step in a much larger program. On October 3, 1997, the head of Iran's Atomic Energy Agency

(Gholamreza Aghazadeh, the former Oil Minister) announced that Iran would pursue a plan aimed at meeting 20% of the country's electricity demand through nuclear power. Aghazadeh said that the government had decided to build a second 1,000-megawatt (MW) unit at the Bushehr nuclear power complex as soon as work is completed on the current unit being built by the Russians. Aghazadeh further said that Iran was discussing further nuclear power plant deals with Russia and China. Since that time, the Khatami government has made similar statements, and has discussed the purchase of two more reactors from Russia. These discussions continued in Moscow in December 1998. It has also discussed the purchase of up to three more reactors from China.

Iran claims that its nuclear power is for peaceful purposes and that it will help free up oil and gas resources for export, thus generating additional hard-currency revenues. On February 23, 1998, the U.S. State Department reaffirmed U.S. opposition to Iran's nuclear program. The United States has argued that Iran has sufficient oil and gas reserves for power generation, and that nuclear reactors are expensive, unnecessary, and could be used for military purposes. Iran has responded that it is a signatory to the Nuclear Non-Proliferation Treaty, complies with all IAEA requirements and inspection procedures, and has no plans to acquire nuclear weapons. President Khatami and other moderate Iranian leaders repeated these statements on numerous occasions in 1998 and 1999.

There are some U.S. and European experts that question whether the light water reactor being constructed at Bushehr is really useful for the production of fissile material, and how much technology transfer Iran can gain. There is strong intelligence evidence, however, that Iran's nuclear reactor program is closely linked to its nuclear weapons program. U.S. policy opposing the reactor is unlikely to change and is likely to affect all aspects of U.S. political and commercial relations with Iran.

Energy Development in Iraq

It is even more difficult to predict the short and mid-term developments in Iraq's energy sector than it is for Iran. UN sanctions remain a critical issue, as do UN controls over Iraq's ability to use export income to invest in its energy sector and import the necessary equipment. There is a continuing risk of war and/or new sanctions, and even of conflict involving the use of weapons of mass destruction. Ironically, Iraq's current Oil Minister has been deeply involved in both its development of weapons of mass destruction and in the controversies surrounding UNSCOM and UN inspections.

The Impact of the Gulf War

The Gulf War has led to fundamental changes in Iraq's energy policies. The war severely hurt Iraq's southern oil industry. The U.S. Department of Energy estimates that capacity fell from 2.25 million barrels per day to 75,000 barrels per day in mid-1991. The largest producing oil field in the region is Rumaila. The Gulf War *destroyed* production infrastructure (gathering centers, compression/degassing stations) at Rumaila, storage facilities, the 1.6-MMBD (pre-war capacity) Mina al-Bakr export terminal, and pumping stations along the 1.4-MMBD (pre-war capacity) Iraqi Strategic Pipeline. SOC has 7 other sizable fields that remain damaged or partially mothballed. These include Zubair, Luhais, Suba, Buzurgan, Abu Ghirab, and Fauqi.

The situation in the north was less serious, but still presented problems. The Kirkuk field, which was originally brought online by IPC in 1934, still forms the basis for northern Iraqi oil production. Kirkuk has over 10 billion barrels of remaining proven oil reserves. The Jambur, Bai Hassan, and Khabbaz fields are the only other currently- producing oil fields in northern Iraq. While much of Iraq's northern oil industry was not affected by the Iran-Iraq War, the Gulf War damaged an estimated 60% of Northern Oil Company's (NOC) facilities in northern and central Iraq. Post-1991 fighting between Kurdish and Iraqi forces in northern Iraq resulted in temporary sabotage of the Kirkuk field's facilities. Most war-related damage was repaired by 1993, although questions remain about water production problems. Since that time, NOC has used field rotation to keep production facilities in working order.

The Importance of UN Sanctions

UN Sanctions have placed severe limits on Iraq's ability to capitalize on its oil resources. Prior to its invasion of Kuwait in August 1990, Iraqi oil production had just recovered from the costly Iran-Iraq War. In July 1990, Iraqi crude oil output had reached 3.5 MMBD, with production capacity estimated at 4.5 MMBD—the highest levels since 1979.

Following Iraq's invasion of Kuwait and the embargo on Iraqi oil exports, oil production fell to only 300,000 barrels per day. Iraq produced an average of 2.04 million barrels a day in 1990, 305,000 barrels per day in 1991, 425,000 in 1992, 512,000 in 1993, 553,000 in 1994, 560,000 million in 1995, and 579,000 in 1996.

About 550,000 barrels per day of Iraq's oil output was consumed domestically, either for refining or for direct burning by industrial customers or utilities. The rest of Iraq's oil exports were subject to tight UN controls which were not eased between 1991 and 1995 because the Iraqi government refused to accept UN controls over its oil export earnings. This situation changed after UN Resolution 986 was passed in April 1995. The resolution allowed Iraq to sell specified amounts of crude oil over six-month periods, although. much of the revenue from such sales was allocated for the purchase of humanitarian supplies for distribution in Iraq under UN supervision. The remaining proceeds were also allocated to pay compensation for Gulf War victims, pipeline transit fees for Turkey, and fund the UN special commission (UNSCOM) that was attempting to dismantle Iraq's capability to produce weapons of mass destruction.

In 1997, Iraqi oil production still averaged only 1.19 million barrels per day, and Iraq averaged only 650,000 barrels per day (bpd) *in net exports* of crude oil. An estimated 57% of this oil flowed through the Iraq-Turkey pipeline, conforming to the UN Resolution 986 mandate that at least half of the "oil-for-food" exports must transit through Turkey. The United Nations authorized Iraqi export of oil to Jordan as an exception to the general embargo on Iraqi oil exports. In December 1997, Iraq agreed to increase its crude oil and refined product exports (at half the world price) to Jordan in 1998 to 96,000 barrels per day, from about 90,000 barrels per day in 1997. Iraq uses revenues from these sales to buy Jordanian goods and services and to settle outstanding Iraqi debts to Jordan.

In addition to UN-sanctioned oil exports to Jordan, Iraq smuggled up to 100,000 barrels per day of crude oil and products to Turkey via truck, to India and Pakistan along the Gulf coast from Jebel Ali, to Iran across the Fao Peninsula with barges, and to Dubai with the use of small tankers sailing from Umm Qasr. Press reports also have estimated that these illegal shipments may provide Iraq with as much as $700 million a year in revenues.

The Impact of the UN "Oil for Food" Program

The reason for the rise in Iraqi oil production and exports was that Iraq accepted the terms of UNSCR 986 in May, 1996 — 13 months after the

UN first made the offer. Iraq began to implement the program in December 1996, and the first shipments paid for by oil exports arrived in March 1997. More than 7.5 million tons of food, about $400 million of medicine and medical supplies, and more than $200 million of agriculture, electricity, water, sanitation and education supplies and equipment arrived in Iraq.

On January 20, 1998, the UN Security Council voted unanimously to more than double (to $5.26 billion) the amount of oil Iraq could sell over a six month period. Exports in February 1998 rose to 1.2 MMBD (about 60% Kirkuk and 40% Basrah Light). Of this, about half was destined for European markets, about 40% for the United States, and 8% for Asia. It is far from clear, however, when Iraq will be able produce enough oil to reach the new UN limit.

According to a variety of estimates, Iraq's crude export capacity was then somewhere between 1.4 and 2.4 MMBD (a recent Dow Jones survey of oil analysts settled on 1.7 MMBD). This included 0.8-1.6 MMBD (Iraq estimates 1.2 MMBD) through the Kirkuk-Ceyhan pipeline, and 0.6-0.8 MMBD through Mina al-Bakr. This was a far smaller total export capacity than is necessary to sell the $5.26 billion worth of oil which it is authorized to do under the UN-approved plan. Even at $13 per barrel for Iraqi oil, Iraq would have had to export about 2.2 MMBD over 180 days to reach $5.26 billion — far beyond the country's stated current capacity.

Iraq announced in early September, 1998, that it had exported 20 million barrels of oil from its southern Mina-al-Bakr terminal in June under the oil-for-food deal with the United Nations. The head of the Iraqi Southern Oil Company, Mohammed Yassin al-Douboni, said that the capacity of the export terminal, which was bombed during the 1991 Gulf War, had been raised to 1.8 million barrels per day (barrels per day).

A few days later, however, the UN reported that Iraq could produce only a little more than half of the $5.2 billion in oil it *was* allowed to sell over the coming six months. Iraq was authorized by the Security Council to sell $5.2 billion worth of oil during the previous six months, but it *had* only managed to sell $1.06 billion worth of oil because of weak world demand and the damaged state of its oil industry. Because of weak world oil prices, Iraq was getting an average price of only $9 to $10 a barrel for its oil compared with $19 to $20 when it first started selling oil under the program in December 1996. As a result, oil experts warned that the country was unlikely to be able to raise more than $2.5 billion to $3 billion during the current six month period.

The UN reported on November 17, 1998, that Iraq shipped 286.8 million barrels in the fourth six-month phase of the program, and had commitments to ship another 21.655 million barrels in the 12 days from November 14 to November 25. So far, it had shipped $2.81 billion of oil in the oil sale's fourth six month phase. This sales level was expected to be between $3.0 billion and $3.1 billion by November 25, well short of the $5.256 billion Iraq was allowed to sell, although far more than the $2 billion that Iraq has been allowed to earn during each of the three previous six month periods since May 1996.

Iraq's production rose during the months that followed. The UN announced on December 28, 1998, that that Iraqi oil exports had reached a record 2.51 million barrels per day during mid December, in spite of ongoing U.S. and British air and missile strikes during Desert Fox. Furthermore, eight contracts for the sale of oil were approved in the week from December 19-25.

This brought the number of approved contracts for the six month period to 58, for a total of 224.2 million barrels of oil.

Unfortunately for Iraq, its average total production for 1998 was still only 2.1 million barrels per day, and increased production did not mean increased revenue. The price of Iraqi oil dropped to a new low amounting to $8.23 per barrel. However, the UN estimated that the increase in oil prices allowed Iraq to earn more than $3.7 billion during the six months ending on May 24, 1999. Ironically, every increase in Iraqi oil production was also helping to increase the same oil glut that was creating low oil prices. The UN exports were also concerned with the fact Iraq might be overproducing and permanently damaging some of its oil fields. They warned that if Iraq continued to ship oil at this relatively high volume it might damage its oil fields, and that they need extensive rehabilitation, and gas injection, to maintain the production of Kirkuk and Basrah Light crude oil, which fell to 1.87 MMBD in June 1999.

In May 1999, Secretary General Kofi Annan issued a report on the operation of the oil for food program from its inception in December 1996 until November 1998. It recommended that spare parts for Iraq's oil industry be funded outside of the program, which currently earmarks $300 million for that purpose every six months. This would allow for $300 million more for humanitarian purposes while still allowing Iraq access to the spare parts it needs to maintain and increase production at its oil fields.

The Routes for Iraqi Oil Exports

Iraq is experiencing other problems because it is shipping more oil from Mina al-Bakr, its terminal in the Gulf, than from Ceyhan, Turkey. The Security Council had required that the majority of the oil be shipped from Ceyhan. This is to allow Turkey to gain pipeline fees that it has lost since Iraq invaded Kuwait in August 1990 and the Kirkuk-to-Yumurtalik pipeline was shut. Through November 13, 1998, a total of 153.705 million barrels had been shipped from the Turkish port and 154.750 million barrels had been shipped from Mina al-Bakr. It shipped Basrah Light crude from the Iraqi port and Kirkuk crude from the Turkish port.

These problems may become even more severe if Iraq continues its efforts to reach out to some of its former Arab enemies. In August 1998, Iraqi Oil Minister Amir Muhammed Rasheed agreed with his Syrian counterpart on a schedule to pump oil via the pipeline, closed since 1982.

Officials from the two countries say steps are under way to reopen their embassies, closed for the past two decades.

Iraq has also agreed to set up a pipeline with Jordan which would cost $350 million, and extend the existing pipeline from Hditha, 260 kilometers northwest of Baghdad, by 750 kilometers and reach Jordan's refinery in Zarqa, northeast of Amman. Jordan currently imports 4.5 million tons of crude and product a year from Iraq – roughly 75,000 barrels a day – under terms authorized by the UN. Before the Gulf War, Iraq and Jordan had discussed a major pipeline link to Jordan's port of Aqaba on the Red Sea and the joint construction of a 350,000 barrel per day refinery.

Iraq's Deputy Oil Minister Taha Hamoud announced in September 1998 that Iraq was ready to export oil under its oil-for-food deal with the UN via the Iraqi-Saudi pipeline, which had been closed since the 1991 Gulf War: "We are ready to pump oil via the pipeline which connects Iraq with the Red Sea with an export capacity of 1,650 million barrels per day. " Iraq's Minister of Trade Mohammed Mehdi Saleh says Iraq is holding talks with Saudi firms to buy food and medicine under its oil-for-food deal with the United Nations. ``We are conducting trade negotiations with our brothers the Saudis to conclude food and medicine contracts ...Representatives of several Saudi companies have visited Baghdad recently and we are currently holding talks with these companies.

Investing in Rehabilitating Iraqi Oilfields

Another debate has taken place over Iraq's effort to import equipment to repair and expand its oil production capability. In June 1998, the UN approved the sale of up to $300 million in oil infrastructure equipment to be funded through Iraqi oil sales. In reviewing the resulting Iraqi requests for purchases of parts and equipment, the United States has distinguished between 'upstream' equipment — getting the oil out of the 'downstream' operations, which involve getting petroleum products to market. U.S. officials have said that the $300 million approved by the Security Council for spare parts should only be used for short-term improvements and not to overhaul Iraq's oil industry for its long-term needs.

This has led to *increased* friction between the U.S. and Iraq, and delays in the program. On September 6, 1998, Benon Sevan, UN director of the oil-for-food plan criticized the Security Council, especially the United States, for holding up equipment Iraq needs to upgrade its oil industry, one reason the so-called oil-for-food program was falling far below targets. "The program cannot be fully implemented unless the Iraqis can enhance production and enhance capacity to produce oil ...This is a crucial issue which needs urgent resolution. " Sevan said the delay in obtaining oil industry equipment had led Iraq to tell purchasers it would have to reduce oil supply contracts by some 10 percent over the next three months

Sevan said that oil experts from the Dutch Saybolt firm made it clear to the Security Council that downstream operations were needed to enable the industry as whole to operate safely and efficiently. He said he had discussed the issue with U.S. officials and "I think there is movement, a little bit on that issue. " He also said, however, that the entire program was being run in a 'suspicious' political atmosphere that made progress difficult. He said that 67 equipment contracts were received for approval at a total value of $79.8 million, of which 52 had been circulated to sanctions committee members. A total of 19 were approved and 22 applications had been placed on hold. Sevan said the shortfall in revenues delay a rise in Iraqi food supplies to 2,300 calories a day per person from the current 2,000 calories until at least October, noting that the calorie level was only 1,400 before the program began 18 months ago. Sevan says the quality of foodstuffs was poor and probably could not be improved because of the lower revenues.

In October 1998, Iraq accused the United States of blocking UN approval for 40 contracts it signed with international firms to supply spare parts for its oil industry. Oil Minister Lt. Gen. Amer Mohammed Rashid said that despite the council's approval, the UN sanctions committee has not reached agreement yet on how to process the contracts. He blames Washington for the delay. "We have almost no spare parts now. We are running at a very critical level in terms of production."

Nevertheless, the Sanctions Committee had approved seven spare parts contracts with a value of $4.6 million by the week ending November 13, 1998. The costliest item was a $2.5 million passenger boat from Russia. The UN had received requests from Iraq for 324 "spare parts" contracts worth $181.1 million. Of these, the UN Security Council's Iraq sanctions committee has approved 139 contracts worth $97.5 million and withheld approval for 96 contracts worth $44.9 million.

The purchasing efforts acquired growing momentum towards the end of 1998. The Committee approved 13 more contracts in early December 1998, worth $4.4 million, for spare parts and other equipment Iraq needs to upgrade its dilapidated oil industry. The purchases came from 13 countries, with the largest contracts going to Belgium, the United Arab Emirates and Germany. Baghdad is allowed to import $300 million in spare parts out of its oil revenues and has purchased $133 million worth since June when the council first approved the expenditures. However, the Sanctions Committee did delay approval of another 133 contracts, worth $43.6 million. These contracts had been on hold since June, and six of them were delayed in the second week of December. By January 1999, the UN had approved a total of $261.14 million out of the $300 million available under phase four of the oil-for-food deal. The purchase of a total of $600 million worth of spare parts had been approved since June 1998.

Nevertheless, Iraq was still accusing the U.S. of holding up spare parts for its oil industry in late January 1999. A U.S. official confirmed that the U.S. had put 12 contracts on hold so experts could review them but that the U.S. had released 77 contracts it previously put on hold since January 15, and had a policy of expediting such contracts. He said *that the* U.S. had lifted objections to 43 contracts the previous week, leaving another 114 contracts worth $45 million on hold according to the UN, but a total of 142 contracts according to Iraq. The U.S. had objected to some contracts because they were for parts to repair refineries from which Iraq smuggled oil rather than using it to purchase food, medicine and other goods.

There is some uncertainty about just how serious Iraq's current production problems are. Four experts from Saybolt, the Dutch firm reviewing Iraqi oil field development for the UN, surveyed Iraq's needs in December 1998. And reported that Iraq's production was averaging about 2.5 MMBD, with 550,000 b/d of domestic consumption and shipments of 70,000 b/d to Jordan. The remainder could be exported and Iraq averaged 1.71 MMBD during the last six months of 1998. The Dutch experts found, however that Iraq was losing about 4-8% of its production capacity a year, and that a significant number of wells had ceased production in the north and south because of a lack of water removal, which was producing permanent damage to about 20% of the wells. It reported that the Iraqi Oil Ministry had claimed that 25 wells had been lost in the north, and that 30 more would be lost before the arrival of spare parts. Saybolt also reported that even if Baghdad spent the $600 million it had been permitted, oil production would likely remain at the 2.5 MMBD level. The Saybolt experts also reported, however, that the entire program developed by the Iraqi oil industry did not follow modern engineering practices and did not offer value for money. It felt that many of Iraq's requests for spare parts might be technically sound, but relied on outdated concepts for oil field development. It is not clear how the Iraqi Oil Ministry assessed these comments, but it changed its estimates to indicate that Iraq could at best sustain a maximum production level of 2.5 MMBD until the projected start up of major new water injection activity in March 2000. Iraq has tried to deal with the problem of managing the development of individual parts of its oil industry by crreating thirteen independent operating companies, which were set up in March 1998. These "companies" all now have individual budgets for imports of spare parts:

- North Oil Company ($85 million): Some 1.2 MMBD of production capacity in the north. This is the company which claims to have lost 25 wells and face the loss of 30 more. It needs extensive dewatering activity.
- South Oil Company ($100 million): This company is in charge of the North Rumaila oil field re-development project. It needs new water injectors and control equipment, and has the goal of increasing production by 200,000 b/d.
- Iraqi Drilling Company ($20 million): Iraq's existing drilling rigs are in a "lamentable condition" according to Saybolt. Extensive drilling is needed for oil and water.

- Oil Projects Company ($10 million): A major construction contractor.
- Oil Exploration Company ($5 million): Tasked with examining the amount of oil and water in given resevoirs.
- North Refineries Company ($32 million): Needs funds for a new 60,000 b/d hydrocracker, spare parts for the Baji refinery, and other repairs.
- Midland Refineries Company ($10 million): Boost output at the Daura refinery at Baghdad with new wastewater facility and repairs to lubricating oil plant.
- South Refineries Company ($5 million): Boost production at Basra refinery.
- North Gas Company ($4 million).
- South Gas Company ($4 million).
- Oil Products Distribution Company ($10 million): Vehicles, fuel dispensers, meters, etc. to modernize fuel distribution system.
- Oil Pipelines Company ($5 million).
- Gas Filling Company ($10 million): 500,000 propane tanks, new filling and safety equipment.

The Hardship Issue

Humanitarian issues are involved as well as energy policy. Iraq's oil exports have not earned anything like its quota of $5.26 billion every six months, and the total value of its sales were worth less than $3 billion during the last six months of 1998. *There* are severe limits to how Iraq can allocate the revenue it does receive from "oil for food. " While two thirds of the money raised from oil sales is available for buying food and other humanitarian goods, another 30 percent goes into the Kuwait compensation fund and the remainder is spent on the arms inspectors and administrative expenses. There are also constant problems with the approval of orders and the failure to issue the proper orders, and the Iraqi people continue to suffer.

There is considerable controversy *over* how real the "hardship" issue is, and some on-the-scene observers feel Iraqi is politicizing it and exaggerating the suffering of its people. According to UN figures published in the spring of 1997, Iraq imported 6.5 million tons of food and $330 million worth of medical goods since the first shipments *began* in March 1997. Nevertheless, the UN concluded that it needed to allocate more money to food, and less to other humanitarian needs like medicine and health care, if it was to succeed in its objective of raising the nutritional value of the food ration each Iraqi receives from 2,030 calories a day in 1997 to 2,268 calories this year. (2,000 calories is regarded as about the minimum for healthy living.)

The United Nations Children's Fund warned in September 1997 that unless the new UN oil-for-food plan for Iraq *was* properly implemented, it would do little to offset worsening malnutrition suffered by Iraqi women and children. The UN report claimed *that* the infant mortality rate had nearly doubled between 1990 and 1994, while the number of mothers dying while giving birth rose from 117 to 310 per 100,000 *births* between 1990 and 1996. According to a UNICEF survey, over one million Iraqi children under five are suffering from malnutrition.

UN officials held repeated discussions with Baghdad during 1997 and 1998 asking it to prioritize its humanitarian purchases. Iraq responded by complaining about delays in contract approvals and deliveries of goods. Iraqi health officials have charged that less than 1 percent of the $200 million in medical supplies that Baghdad was permitted to buy during the first six months of 1998 had arrived in the country. UN spokesman Eric Falt acknowledged that medicine deliveries were "slower than expected,'' but said $17 million, or 9 percent of the medical supplies, *had* arrived during the third phase.

In spite of a steady improvement in the flow of goods during 1998, Health Minister Umeed Madhat Mubarak claimed in August that more than one million Iraqi children had died as a result of sanctions. He said an average of 6,452 children under the age of five were now dying each month, compared with 539 a month before the Gulf crisis.

The Iraqi Health Ministry claimed on December 29, 1998, that more than 8,800 Iraqis had died in November as a result of UN sanctions and that, "6,269 children below five years and 2,584 elderly persons died during November 1998...as a result of different sorts of diseases caused by the continuation of the embargo imposed on Iraq more than eight years ago. " The Ministry said its data showed that in November 1989, the year before the imposition of sanctions, only 258 Iraqi children and 422 adults died, and that its statistics for children's deaths in November 1998 revealed that 1,631 died *from complications of* diarrhea, 2,419 of pneumonia and 2,219 of malnutrition. Of the adult deaths, 579 were due to heart diseases and high blood pressure, 413 to diabetes and 1,592 to ``tumor diseases. "

The Iraq people have suffered, but there are three critical problems with the kind of dramatic claims made by Iraqi officials, and some UN organizations and NGOs.

- First, they almost always ignore the sharp and steady drop in Iraqi living standards after 1983 that occurred because of the Iran-Iraq War, and that the real per capita income of Iraq dropped by over 60% between 1982 and 1989 because of the Iran-Iraq War, and before the Gulf War and UN Sanctions began.
- Second, the Iraqi and UN data usually assume that Iraqi figures for pre-sanctions health, education, and nutrition data are correct – which is little more than mindless rubbish given the impact of the Iran-Iraq War and Iraq's obsessive propaganda during that war.
- Third, Iraqi official census figures show that the population of Iraq rose from 16.3 million in 1987 to 22 million in 1997. If such figures are taken seriously, Iraqi has had the highest rate of population growth in its history during a period of extreme hardship and high infant mortality, and would have to have sustained one of the highest population growth rates in the world for over eight years if its mortality statistics are correct.

The Future of Iraq's Oil Production and Exports

Serious as these short-term issues are, they are only part of the story. Virtually all experts agree that future increases in Iraqi oil exports will have a major impact on the overall patterns in world oil exports. The U.S. Department of Energy estimates that Iraq will increase its production from 2.2 million barrels per day in 1990, and 1.6 million barrels per day in 1997, to 2.8 (2.3-2.9) million barrels per day in 2000, 3.2 (2.8-3.5) million barrels per day in 2005, 3.8 (3.2-4.4) million barrels per day in 2010, 4.7 (3.7-5.4) million barrels per day in 2015, and 5.9 (4.7-7.2) million barrels per day in 2020.

According to the most recent estimates by the Department of Energy, Iraq contains 112 billion barrels of proven oil reserves, along with roughly 215 billion barrels of probable and possible resources. Iraq's oil resources are the world's second largest, exceeded only by Saudi Arabia's.

Iraq's proven oil reserves are located in 73 structures, including at least 6 super-giant, 17 giant, and 20 large fields. Only 15 of these 73 structures have been developed, and currently producing fields contain about 40% of total proven reserves. Iraq's true resource potential may be understated, as recent advances in horizontal/multilateral drilling and enhanced oil recovery likely will increase recovery rates and raise total recoverable reserves. In addition, most of the exploration and production activity in Iraq to date has occurred at the Cretaceous level. Deeper oil-bearing formations at the Jurassic and Triassic levels (mainly in the Western Desert) could yield additional resources, but have not *yet* been explored.

Iraqi oil reserves vary widely in quality, with API gravities in the 24° to 42° range. *Most of Iraq's crude oil used for export comes* from the country's two largest active fields: Rumaila and Kirkuk. Following post-UN sanction development of South Rumaila's Yamamah formation, Iraq plans to export up to 800,000 barrels per day of "Basrah Light" (35°-40° API). There is evidence, however, that capacity expansion at the main Kirkuk and Basrah fields could be limited by the presence of water in the reservoirs, and the consequent need for de-emulsifying chemicals and other dewatering equipment. Also, the Basrah fields may require expensive enhanced oil recovery (EOR) techniques and other infrastructure repairs.

Investment and Saddam

Aside from the present debate over how to spend some $900 million to repair Iraq's oilfields and petroleum industry, Iraq will need massive foreign help to exploit its oil reserves. Basic utilities and transportation are currently high priorities, but the improvement of oil production and export capabilities is the highest single priority because it is the government's chief source of income. Iraq must obtain both foreign investment and technical aid if it is to develop its oil reserves and expand its mid and long-term production. Such investment will involve serious political issues, regardless of the future of sanctions.

The individual contracts for Iraqi energy investment projects may be negotiated through the Iraqi Oil Ministry, but such contracts are highly political and any foreign company will have to deal with Iraqi politics at the level of Saddam Hussein. Nothing significant happens in Iraq without Saddam's permission, and direct or indirect bribery often accompanies the overt contracting and negotiation process. All meaningful negotiation occurs at the highest levels in Baghdad, and any dealings with any U.S. firm will be part of Iraq's political strategy. The Oil Minister is generally a senior political figure as well as a technocrat, with close ties to Saddam Hussein. The present Oil Minister — Lt. Gen. Amer Mohammed Rashid — has some links to Iraq chemical and biological weapons program. Deputy Oil Minister Taha Hamoud is a senior official in the Ba'ath Party, and may be associated with efforts to smuggle in dual-use technology.

The sensitivities involved are indicated by the fact that Iraq has repeatedly accused the United States of blocking the development of its oil industry. On July 8, 1998, for example, it accused the U.S. of blocking UN approval for 40 contracts it signed with international firms to supply spare parts for its oil industry. The UN Security Council had agreed in June to allow Iraq to purchase $300 million in spare parts and equipment to boost oil production. The Oil Minister blamed Washington for the delay. "The U.S. administration hinders anything positive that will be of benefit to the Iraqi people ...We have almost no spare parts now ... We are running at a very critical level in terms of production.'' *Similar complaints have consistently been made since. Iraq contends that the U.S. maintains an aggressive policy aimed at blocking or suspending oil industry contracts.*

The Scale of Oil Production Investment

Iraq is seeking foreign assistance for a second-phase development of the Saddam field, which would raise oil production capacity, as well as 300 Mmcf/d of gas. In May 1997, Faleh al-Khayat, Director General for Planning at the Iraqi Oil Ministry, stated that 3 MMBD of production capacity could be reached within 1 year, 3.5 MMBD within 3-5 years, and 6 MMBD in less than a decade after the lifting of UN sanctions. This was to be accomplished by a 3-phased development effort including: 1) re-working and upgrading existing upstream and downstream facilities; 2) attracting foreign investment for new field development and production; and 3) actively conducting exploration and development activities in prospective areas such as the Western Desert.

Field development work under the three phases must be extensive. The U.S. Department of Energy reports that 33 fields — containing 50 billion barrels of reserves and a potential production capability of 4.65 MMBD — are slated for eventual development. Although development costs in Iraq are as low as $1/barrel, Iraq fully understands that any post-sanction oil program will require massive amounts of foreign investment.

In May 1997, former Iraqi Oil Minister Chalabi estimated that Iraq would need at least $5 billion of foreign investment during the first 2-3 post-sanction years in order to bring the country's oil output back to pre-Gulf War levels. He also projected that $30-$50 billion of foreign investment would be required to bring capacity up to 6 MMBD.

The Role of Foreign Companies

The U.S. Department of Energy estimates that some 60 foreign oil companies from a wide variety of countries have been in discussions with the Iraqi government. U.S. firms that have held talks on Iraqi field development include: Amoco, Arco, Chevron, Coastal, Conoco, Exxon, Mobil, Occidental, and Texaco. Iraq plans to offer new fields to foreign oil companies through production sharing contracts (PSC), joint ventures, and service contracts. Initially, Iraq plans to offer up to 25 new fields to foreign companies. Ten of these fields, with a production potential of 2.7 MMBD, are slated for development under PSCs with foreign companies. Four of these fields are located in southern Iraq and, with a combined production potential of 2.1 MMBD, represent the cornerstone of Iraq's post-sanction development plans. These four "giant" southern fields are Majnoon, West Qurna, Nahr Umar, and Halfaya.

Iraq has signed PSCs (reportedly on relatively generous terms) for two post-sanction field developments. The first agreement is a $1.2 billion contract with the China National Petroleum Corporation (CNPC) and Chinese state-owned Norinco for development of the al-Ahdab field. Al-Ahdab is located about 40 miles south of al-Kut in central Iraq. The field contains an estimated 1.4 billion barrels of oil and has a production potential of roughly 90,000 barrels per day. CNPC and Norinco reportedly have formed a new company, named al-Waha, to undertake the field development. Development and operating costs are expected to be around $1.3 billion.

In March 1997, Iraq signed a PSC with a consortium of Russian firms for second-phase development of the 15 billion barrel West Qurna field, located west of Basra near the Rumaila field. The Russian consortium comprises Lukoil (52.5%), Zarubazhneft (11.25%), Machinoimport (11.25%), and an Iraqi company selected by the government (25%). The West Qurna PSC included development of the Yamamah and deeper Mishrif reservoirs, which, combined, contain close to 8 billion barrels of light (37° API) and heavy (27° API) crude oil. West Qurna has a production potential of 500,000-750,000 barrels per day, two-thirds of which will be heavier Mishrif crude. Lukoil reportedly had been in discussions concerning the West Qurna project since early 1994.

Most of the required production wells have been drilled, although a crude pipeline spur and associated gas processing stations are only partially completed. Completion of first phase development could take up to another year, but it is unclear exactly how much surface work remains and whether this is included in the recent PSC, which is valued at $3.7 billion.

Besides West Qurna, PSCs for the three other large southern oil fields are in various stages of negotiation. The largest of the four fields is Majnoon, which has reserves of 10-30 billion barrels of 28°-35° API oil. France-based Elf Aquitaine has an agreement with Iraq on development rights for this field. In the past, it was reported that Elf would plan to retain operatorship and a 40% stake in the $3-$4 billion project. Initial output is projected at 300,000 barrels per day, with later development yielding 600,000 barrels per day or more. Ultimate production potential is estimated at up to 2 MMBD.

As with Majnoon, the 6-billion barrel Nahr Umar field was explored and appraised by Braspetro in the mid- to late 1970s. Five wells had been drilled prior to the Iran-Iraq War. France-based Total apparently has all but agreed with Iraq on development of Nahr Umar. Initial output from Nahr Umar is expected to be around 440,000 barrels per day of 42° API crude, but may reach 500,000 barrels per day with more extensive development. The 5-billion barrel Halfaya project is the final large field development in southern Iraq. Italian Agip originally drilled four appraisal wells at Halfaya under a service contract in the 1970s. A variety of companies reportedly have shown interest in the field, which could ultimately yield 200,000-300,000 barrels per day in output.

Smaller fields with under 2 billion barrels in reserves also are receiving interest from foreign oil companies. These fields, along with anticipated maximum production levels, include: Nasiriya (250,000 barrels per day); Khormala (100,000 barrels per day); Hamrin (80,000 barrels per day); and Gharraf (100,000 barrels per day). As of June 1997, Italy's Agip was thought to be close to signing a contract for the Nasiriya development. In the past, Agip officials have stated that they would like to use the field's access to infrastructure as a "reference point" for development of a number of satellite fields. Spain's Repsol also appears to be a strong possibility to develop

Nasiriya. In addition to the 25 new field projects, Iraq plans to offer foreign oil companies service contracts to apply technology to 8 already-producing fields. This will include new reservoir development at the North and South Rumaila, Zubair, Luhais, Subba, Abu Ghirab, Buzurgan, and Fuqa fields. Iraq also will provide incentives to promote exploration in the remote Western Desert. Located near the Saudi and Jordanian borders, Iraq has identified at least 110 prospects from previous seismic work in this region. As of December 1997, Calgary-based Ranger Oil was reported to be in discussions with Iraq on a $350 million deal to develop an oilfield in this area.

Changing Saudi Energy Strategy

Saudi Arabia is the key supplier of world oil imports, and is one of the largest and most powerful states in the Gulf. In 1998, Saudi Arabia had well over 260 billion barrels of proven oil reserves, or one-quarter of the world's total. Saudi Arabia is the pivotal oil exporter in the Gulf, the Middle East, and the world. Saudi Arabia currently has a production capacity of over 10 million barrels a day. It produced at levels well in excess of 8 million barrels from 1991 to the present, and averaged over 8.4 million barrels a day in 1998 – roughly 30% of total OPEC crude production). It is the world's largest producer, and its surplus production capacity allows it to act as a "swing" producer, and play a critical role in ensuring moderate and stable oil prices.

Both the Department of Energy and the International Energy Agency estimate that the growth in Saudi oil production will outstrip the growth in all of the nations in the Former Soviet Union, in spite of major increases in production by the former Soviet republics in the Caspian and Central Asia. The U.S. Department of Energy estimates that Saudi Arabia will increase its production capacity from 8.6 million barrels per day in 1990, and 11.4 million barrels per day in 1997, to 11.1 (11.5-11.2) million barrels per day in 2000, 13.7 (12.4-14.8) million barrels per day in 2005, 14.1 (12.9-17.3) million barrels per day in 2010, 16.2 (13.3-21.9) million barrels per day in 2015, and 20.0 (17.8-27.4) million barrels per day in 2020.

To put the resulting growth in Saudi production capacity in perspective, Saudi Arabia is already equal to all other states in the Gulf combined, and will remain so through 2020 in spite of major increases in production by Iran and Iraq. Saudi production capacity is 160% of all production capacity in the FSU today, and the Department of Energy estimates that it will fall to 152% by 2020. The Department of Energy estimates that Saudi Arabia's production will shift from 14.7% of world production in 1997, to 13.8% in 2000, 14.8% in 2010, and then rise to 17.8% in 2020.

Saudi Arabia Triggers the "Oil Crash"

This immense energy potential does not, however, mean that Saudi Arabia has a strong economy or can afford to develop its energy resources at the level required to meet world demand. In fact, Saudi Arabia was a major factor behind the recent crash in oil prices.

Saudi Arabia has long been a leader in OPEC's production quota decisions, and it was Saudi pressure that led to OPEC's decision of December 1, 1997, to raise the oil production ceiling by 10% — although actual OPEC production increased by less than 2% between November 1997 and February 1998). This decision gave Saudi Arabia its highest oil output quota in more than 15 years. The Kingdom secured a new quota of 8.76 million barrels per day (bpd) at OPEC's ministerial meeting in Jakarta. This quota was 760,000 bpd higher than the quota it had held over the last four years, although Saudi Arabia had recently been pumping between 8.2 million bpd and 8.3 million bpd in the months leading up to the agreement.

Saudi Arabia took this action despite the fact that it meant that OPEC's overall production ceiling rose from 25.03 million bpd to 27.5 million bpd, and the concern of other states in the 11-member group that oil prices would fall because of increased supplies. Saudi experts felt that demand and prices were firm enough so that the benchmark price of Brent crude would average $20 a barrel even with the new OPEC ceilings. Brent was then trading at $19 a barrel, equivalent to around $17.50 for a typical Saudi export barrel.

Saudi motives for this action were clear. Saudi officials stated that the rise in quota was necessary to increase the Kingdom's cash flow and funding for its 1998 budget and five year plan to reduce or eradicate its $4.5 billion budget deficit, and clear contractual debt — some of which stretched back to 1991 — and smooth over any difficulties in loan repayments such as the $4.3 billion loan Saudi Arabia took out in November, 1997 to pay for new aircraft for the country's national airline Saudia. Saudi Arabia also sought funding to pay for new infrastructure projects and military equipment, such as replacements for some 100 aging U.S. F-5 warplanes.

The result, however, was a massive drop in oil revenues whose financial impacts still cannot be estimated. Several factors caused the problem. Asia was Saudi's largest crude export market, accounting for 60 percent of sales, with its main buyers in Japan and South Korea. ($1=3.75 riyals). Asia was already in an economic downturn and slid into a major recession even before the new quota went into effect on January 1, 1998.

Oil prices fell sharply from the $23 range in October 1997, reaching a nine-year low of $12.80 a barrel on the New York Mercantile Exchange in March *1998*. Prices rose to as high as $16.50 a barrel when Saudi Arabia, Mexico and Venezuela agreed to a joint cutback of 1.325 million barrels per day, but then dropped back to levels which often fell to around $10 a barrel in spite of further efforts to limit production.

Saudi Arabia faced massive budget deficit and investment problems in spite of a second cutback in production. Commodity-driven economies are always at high risk, and Saudi Arabia has learned the hard way that there is no reliable way to predict or manipulate the price of oil. Supply and demand projections by OPEC, IEA, and EIA also agree that future average oil prices — and Saudi Arabia's real oil revenues — are highly unlikely to rise back to anything approaching their peak levels for the foreseeable future. Consulting firms like Petrofinance predicted that Saudi Arabia faces budget and current account deficits of 9% until prices rise, and Saudi revenues might drop even further in the near to mid-term if Iraq is allowed to resume full production and/or sanctions on Iran are ended. UN plans then called for Iraq to more than double its oil sales to $10.56 billion annually in 1998, and even the U.S. was *becoming* less interested in enforcing economic sanctions on Iran.

The Cost of the "Oil Shock" to Saudi Arabia

This decline in oil revenues represented a major challenge for the Saudi government since oil export revenues account for nearly 90% of total Saudi export earnings. Aside from the overall oil price decline that affected all world oil producers, Saudi Arabia's difficulties were compounded by the economic crisis in Asia, since Asia accounted for around 60% of Saudi oil sales.

In response to the low oil prices, Saudi Arabia led the way in establishing serious OPEC cuts in production. At its March 23, 1999 meeting, 10 OPEC nations (excluding Iraq) agreed to cut production by 1.716

MMBD in addition to previous cuts. These production cuts were successful in raising oil prices, and by June 1999, OPEC "basket" prices had risen to $15.43 and Arab Light had risen to $15.60. Despite the increase in prices, it is clear that the "oil shock" had a great effect on Saudi Arabia, and one that should not be overlooked.

The Department of Energy estimated that Saudi Arabia earned about $48.1 billion in 1997 from crude oil exports. In 1998, this fell by 38%, to around $29.7 billion (Saudi Arabia loses an estimated $2.7 billion for every $1/bbl fall in the price of oil). After January 1998, the price of Saudi Arab Light averaged between $10 and $13 per barrel, down around $7 per barrel *in comparison to the prices of* the last few months of 1997. These prices were also well below the Saudi estimates *of* $16 per barrel, *which it used in planning its* 1998 budget. In inflation adjusted terms, these represented the lowest Arab Light prices for a sustained period of time since 1973.

Such massive cuts in revenues meant that Saudi Arabia faced major budget deficits in the interim, and was heavily dependent on oil revenues (88% of total export earnings, about 75% of state revenues, and 40% of GDP). The drop in Saudi export revenues also came just as the Saudi economy appeared to be recovering from the adverse impacts of the 1990/91Gulf War, and resulted in a negative rate of GDP growth compared to an average 4.4% annual growth rate over the past two years. They also led to a much higher budget deficit of $ 12.3 billion (9% of GDP) in 1998.

These financial pressures were particularly serious because of the failure of Saudi economic reform. Despite periodic efforts to diversify, Saudi Arabia's economy remained essentially oil-based (although investments in petrochemicals had increased the relative importance of the downstream petroleum sector in recent years). In 1997, 88% of Saudi Arabia's export revenues were derived from the sale of crude oil, natural gas liquids, and refined products.

Saudi Arabia had applied for membership in the World Trade Organization (WTO) as part of a strategy aimed at increasing integration into the world economy. In order to gain acceptance into the WTO, however, Saudi Arabia had to cut tariffs, a step which was complicated by the Saudi commitment to a common Gulf Cooperation Council (GCC) tariff and customs union, which despite 14 years of negotiations are still not in place. As a result, it made slow progress.

The Saudi government five-year plan for 1995-1999 sought to reduce state involvement and increase private sector (including foreign) participation in the economy. The government moved slowly, however, in making economic reforms like (cuts in subsidies, increases in taxes, privatization, financial sector reform), however, because of its concerns over the possible social unrest that such moves could entail. Saudi Arabia also adopted a policy known as "Saudiazation," the goal of which was to increase employment of its own citizens by replacing 60% of the estimated 5-6 million foreign workers in the country. Saudi Arabia did stop issuing work visas for certain jobs, moved to increase training for Saudi nationals, and set minimum requirement for the hiring of Saudi nationals by private companies. It was not until 1997, however, that Saudi Arabia seemed to be serious in enforcing this policy.

In March 1999, OPEC agreed to a 1.716 MMBD cut in production effective April 1, 1999. Out of this total, Saudi Arabia agreed to cut 585,000 bpd, bringing its quota down to 7.438 MMBD. These cuts, coupled with an unusually high compliance rate (90% in May 1999), were successful in raising oil prices to the $15-$17 range by mid June. However, since higher oil prices were achieved through lower production, it is unlikely that any of the OPEC nations, including Saudi Arabia, will experience significantly higher oil revenues in 1999. In fact, the EIA estimates that Saudi Arabia's oil revenues will only increase 1% in 1999 from 29.7 billion to30.1 billion, though this estimate might be a little conservative.

Given this background, Saudi Arabia can seek to cope with lower oil short-term energy export revenues in a number of ways, although each option has its drawbacks in practice:

- Saudi Arabia can attempt to make further cuts in oil production, hoping that this would raise marginal prices to the point where it would result in a significant increase in real export revenues. It is unclear that this strategy can do anything other than make things worse, however, because other producers are likely to take Saudi Arabia's market share to maximize their own revenues.
- Saudi *Arabia* can reduce its domestic subsidies (on gasoline, electricity, and water, for instance). Reduction in popular entitlement programs, particularly in areas like health care, housing, jobs, and education, however, could risk political or social unrest.

- Another possibility is to draw down the country's large foreign exchange reserves (around $8 billion). Saudi Arabia also could increase borrowing and/or sell off state-owned (including foreign) assets, although this risks exhausting the country's once huge financial reserves.
- Saudi Arabia can cut back on defense spending, which accounts for an estimated one-third of the Saudi budget. Cuts in defense, however, are constrained by perceived and actual threats from neighboring countries like Iraq, as well as the need for domestic security.
- Other possibilities for budget cuts include investments in the oil and gas sectors, cuts which Saudi Aramco has already made to some extent. Saudi Arabia's Ministry of Finance and National Economy is pressuring ministries to cut expenditures, especially in their capital budgets. Reductions in wages and salaries for state workers are possible, but less likely, since they would undoubtedly be politically unpopular.
- Finally, Saudi Arabia can raise taxes, although this would also be politically unpopular.

Changes in Saudi Energy Investment Policies

There are some experts who argue that the decline in oil prices and cutbacks in production may serve Saudi Arabia's long-term interests by deterring development of alternative energy sources, including increasingly economical "unconventional" oil sources such as Canadian tar sands and Venezuelan orimulsion; by maintaining Saudi share of the market against its main competitors both in and out of OPEC, especially in the key U.S. market; and by deterring marginal non-OPEC oil production investment. This kind of reasoning, however, provides little consolation to a nation that has ahd to make major production cutbacks and which must deal with what is likely to be at least half a decade of severe financial and cash flow problems. It also means that Saudi Arabia must rethink both its investment policies and mid-to-long term energy strategy

It is clear that Saudi Arabia will need increased foreign investment in many of its energy sectors almost regardless of the energy strategy it pursues. Saudi Arabia has recently indicated that it is considering changes in its energy policies which would open up the national oil industry to foreign investment and seek relatively rapid increases in oil production capacity that would ensure that Saudi Arabia remains a dominant swing producer. Crown

Prince Abdullah met with several major oil companies *during his* September 26, 1998 visit *to* Washington. The companies included Mobil, Texaco, Exxon, Chevron, Arco, Phillips Petroleum, and Conoco, and invited them to help develop Saudi Arabia's massive oil resources. He called for proposals by November 10, 1998. The details of the Saudi proposals and the company responses have not yet been released.

A considerable debate then began in Saudi Arabia over what kind of energy development strategy the Kingdom should pursue and what kind of foreign investment is desirable.The incentives for private and foreign investment include Saudi Arabia's massive oil reserves and low extraction costs. The incentives to Saudi Arabia for moving away from a reliance on the state sector are that Saudi Arabia has had more than 10 years of budget deficits. In 1998, the Kingdom's oil revenues had dropped 60% from their level in 1996, and Saudi Arabia faced a budget deficit that could reach up to $8 billion. There were rumors in October, 1998 that Saudi Arabia had to arrange a *$5 billion* private loan from Abu Dhabi.

Other incentives include the need for mid to long-term capital to fund massive civil infrastructure costs, and the need to modernize Aramco's operating procedures and technology base. While Aramco is one of the most efficient State-owned entities in the world, it has suffered from a lack of cash flow for exploration and development in recent years, and efforts at Saudiazation have created problems because skilled foreign managers and technicians have been replaced with Saudis who lack their background and experience.

The key question now is what part of its energy sector Saudi Arabia will open to foreign investment, and the current debate focuses on whether this investment should include downstream oil production, or simply gas production and upstream oil products. The views of Abdullah Dabbagh, a member of Saudi Arabia's consultative Shura Council, and a member of the council's economic and finance committee provide some insights into Saudi attitudes on the question. Dabbagh is also chief adviser to the board of the Council of Saudi Chambers of Commerce and Industry, which he headed for 15 years and is influential in Saudi business circles.

Dabbagh gave an interview on November 18, 1998, on foreign investment in Saudi Arabia's strategic upstream energy sector. Dabbagh said that he believed the Kingdom would consider opening its gasfields to foreign investors. "I would say it's a fair possibility that the government would look seriously at suggestions of investing in upstream gas projects. Gas will be the limit because (investment in) upstream oil is out of the question.

Dabbagh said he would "definitely" support the opening of the upstream gas sector to foreign investors. Dabbagh predicted that low oil prices, the need for advanced technology, and rising Saudi electricity demand would create opportunities in upstream gas. "For the time being, with the decline in oil income, and so on, it might be very hard for us to go it alone. The technology in gas has changed a great deal in the last 10 years or so. Gas is very important. We have to generate a great deal of electricity in the next 20 years. " Dabbagh said Saudi businessmen had told him they would seek joint venture opportunities if foreign investors were given access to upstream gas projects. "A lot of businessmen would be interested in that. I am talking about deep-pocket businessmen. The majority of businessmen would like to see the country open up."

Dabbagh also, however, made comments that reflected the fact that some senior Saudis do not support opening up the upstream sector of Saudi oil production to foreign firms. He said that Prince Abdullah had merely asked U.S. executives for ideas on developing Saudi Arabia's energy sector. "The prince opened the door for ideas. He did not commit himself to anything. We are rational enough to accept good ideas if they come along. " Dabbagh said that it would not make sense to allow foreign investment in upstream oil projects in a country that already has more than two million barrels a day of unused capacity. Dabbagh also said that Crown Prince Abdullah had held a question and answer session in early November with members of the Shura Council about his overseas tour, parts of which were televised. However, the Council was not been formally consulted on the issue of foreign investment in the upstream energy sector. Dabbagh's views are similar to those of many insiders, who stress that the Kingdom is interested in foreign investment in areas like gas exploration and production, and the production of gas-to-liquids and superlubricants, rather than upstream oil operations.

The U.S. government and international oil companies have an obvious interest in opening up downstream oil production, as well as gas and upstream manufacture of oil products. On November 17, 1997, U.S. Energy Secretary Bill Richardson issued a statement publicly supporting an expansion of U.S. energy investment in Saudi Arabia, and announced a visit to Saudi Arabia in January, 1998 to encourage U.S. investment contacts.

Richardson suggested —despite Saudi signals to the contrary — that this might involve the country's prized upstream exploration and production sector.

Richardson told a news conference he understood some form of involvement in Saudi Arabia's upstream was possible, "My understanding is that Saudi leaders want to see upstream investment and joint projects. They haven't been specific entirely yet. Saudi Arabia's interest in opening its upstream market has been widely reported — hailed by industry as a potential major breakthrough. The U.S. government views this as a healthy development. We will encourage it and that will be one of the main purposes of my trip, to encourage this joint upstream investment in both Saudi Arabia and Kuwait. " Richardson also said he had discussed Kuwait's plans in a meeting in the United States last week with Kuwaiti Oil Minister Sheikh Saud Nasser al-Sabah.

When Richardson visited the Kingdom in early February 1999, however, he seems to have been told that the Saudis wished to avoid any foreign investment in oil production. Saudi Arabia's oil minister, Ali al-Naimi, publicly invited U.S. oil companies to submit proposals to utilize the kingdom's natural gas reserves for industrial development, while ruling out any access to its oil reserves, the largest in the world for the time being.

Ali al-Naimi, spoke at a press conference in Riyadh after his meeting with U.S. Energy Secretary Bill Richardson, and said "a successful proposal would be the one on the development of our industrial base being fueled by natural gas. " He added, that "those companies who invest today to develop the industrial base of Saudi will probably be the ones that will participate in the exploitation of our upstream oil if and when it opens up. We are looking for concepts and proposals that deal with integrated projects, which as far as gas is concerned will give us an end-product such as desalinated water, megawatts for increased power, and petrochemicals. In the area of oil we are looking at improving the efficiencies of our refineries, and we are looking for an improved production of lubricants …reasonable people do reasonable things — we have 261 billion barrels of oil reserves, we have over 80 oil fields and we are only producing from 8 or 9, we have 2 million barrels of oil per day of extra capacity — we do not need to expand our oil capacity right now …"

Saudi officials also raised the issue of foreign investment in electricity during Secretary Richardson's visit. They noted that Saudi Arabia needed about $120 billion for power generation projects over the next 20 years, with the annual demand growth for electricity in the Kingdom estimated at 4.5 percent. To meet this demand the country must increase its power generation capacity to 70,000 megawatts by the year 2020, from 21,000 megawatts at present, he said.

Saudi Arabia may well be forced to turn to foreign firms for downstream investments in the future, but a number of international oil companies have made it clear that they are interested in the projects the Kingdom already is opening up to foreign investment. Chevron Chief Executive Kenneth Derr said that his firm submitted plans for the development of all areas of Saudi Arabia's energy industry and that said he was "very encouraged'' by his meetings with Saudi oil officials. In December, Texaco Inc. Chairman Peter Bijur had flown *to* Saudi Arabia to present a natural gas development proposal. Atlantic Richfield Co., Conoco Inc. and Phillips Petroleum Co. have all announced they plan to submit project proposals to the Saudis. Exxon and Mobil also said they are interested in exploring for and extracting oil in Saudi Arabia. Ali-Naimi stated that, "So far we have received concepts, not structural proposals ...several of the concepts we have received in gas is exactly what we are looking for. " Saudi oil ministry officials said that these proposals could become contracts "before the end of the year."

Interestingly enough, some European companies have warned against the massive early development of Saudi oil resources on the grounds they could lead to a new supply and demand crisis. Franco Bernabe, managing director of Italian energy giant ENI, a company that has actively sought closer involvement in Saudi Arabia, said such a development could lead to sharp price rises, especially at a time when global reserve replacement lagged consumption growth: "We will see a period of low oil prices, but then this period will trigger a new oil crisis."

Saudi Oil Production

Saudi Arabia's energy strategy is critical because the Kingdom is the world's leading oil producer, exporter, and holder of spare oil production capacity. Saudi Arabia is a key oil supplier for the United States, Europe, and Japan, although Western Hemisphere producers (Venezuela, Canada, and Mexico) have challenged Saudi Arabia's dominance in the U.S. market in recent years. The Asian market now takes about 60% of Saudi Arabia's crude oil exports, as well as the majority of its petroleum product exports. Europe is Saudi Arabia's second largest oil export market, followed by the United States.

There are three basic energy strategies that Saudi Arabia can pursue:

- Its current strategy of responding to market forces by increasing oil production capacity to try to maintain both moderate prices and a high output of oil to maximize oil revenues. This strategy calls for major increases in oil production for exports, and in gas to meet domestic private and industrial consumption. It also calls for growing Saudi investment in downstream operations and product output to maximize value added sales from oil.
- Leading restraint in production in an effort to maintain high prices and maximize oil revenues. This strategy depends, however, on the Kingdom being able to persuade enough competing exporters to limit production to levels which will produce more of a marginal increase in oil prices over time than normal market forces, and on the ability to achieve a new equilibrium between supply and demand so that the net real oil revenues to the Kingdom are actually higher than maximizing production.
- Maximizing production capacity to recapture market share and marginalize high cost producers. This strategy would produce limited near term oil revenues, but it would potentially change the oil industry and drive out higher cost producers, increasing Saudi market share and net Saudi oil revenues.

It is important to note that the limited production strategy has to some extent already failed. Two production cuts during 1998 did not stabilize prices, and work by the Petroleum Finance Corporation indicates that Saudi Arabia would have had to assume at least 50% of a 1.5 MMBD additional cut in OPEC output to drive prices back to $17 a barrel. This would mean bringing Saudi oil production down to levels of around 7.25 MMDB. It is uncertain that such a strategy can be implemented at all. Iran is desperate to produce and maximize cash flow. Iraq has already stated that it will not restrain production, and a number of other nations continue to cheat on their production quotas. A cartel-like strategy seems to be doomed before it begins. This study by the Petroleum Finance Corporation also indicated that such a strategy would only produce tenuous short-term benefits if it did succeed, and would still not relieve the Kingdom's problems in cutting its budgets, implementing economic reform, and obtaining foreign investment. Such a strategy might reduce the 1999 budget deficit from 44.1 billion Saudi riyals to 31 billion, but it would produce only limited incremental short-term revenues and would probably have negative mid-term impacts because of Saudi Arabia's failure to take sound measures to reform its economy. The end result would be sustained Saudi budget deficits of 36 billion riyals a year.

Saudi production cutbacks in 1999 did raise prices, but the less oil that was sold meant that oil revenues only shifted from $29.7 billion to an estimated $30.1 billion. This was scarcely enough to meet Saudi needs.

The third, or "market share, " strategy is examined in more depth later in this section, but it is important to understand that Saudi Arabia is already planning major mid to long-term increases in production that make it difficult to distinguish between its existing strategy and some aspects of a market share strategy. According to Saudi Oil Minister Ali Naimi, Saudi Arabia's oil policy currently focuses on maximizing revenues, increasing Saudi market share, and maintaining international oil market stability. Of Saudi Arabia's total oil production capacity, about 18% is considered medium-gravity, 14% heavy, and the remaining 68% light. In order to maximize its oil revenues without violating its OPEC quotas, Saudi Arabia has increased Extra Light and Arab Super Light production at the expense of Arab Heavy and Arab Medium crude oil. Although some capacity has been *reduced*, Saudi Arabia is also using field rotations to minimize reservoir pressure damage and preserve the option of bringing the capacity back online within a few months, if needed. Most of Saudi Arabia's spare production capacity is medium crude.

Saudi Arabia produced an average of 6.41 million barrels a day in 1990, 8.1 million in 1991, 8.3 million in 1992, 8.2 million in 1993, 8.1 million in 1994, 8.2 million in 1995, 8.2 million in 1996, 8.6 million in 1997, and 8.4 million in 1998. Saudi oil production totaled about 9.3 million barrels per day in 1997, including 800,000 barrels per day of natural gas liquids and about 270,000 barrels per day of crude oil produced from its half-share of the Saudi-Kuwaiti Neutral Zone. This compared with what the Department of Energy estimates was a sustainable crude oil production capacity of about 11.3 million barrels per day and what was then an OPEC crude oil production quota of 8 million barrels per day (8.76 million barrels per day as of January 1, 1998). Saudi Arabia was producing close to its OPEC quota (production of natural gas liquids is not counted against the OPEC quota), but about two million barrels per day below its capacity.

At OPEC's November 1997 meeting, Saudi Arabia's crude oil production quota was raised about 10%, from 8 million barrels per day to 8.76 million barrels per day. *In* early February 1998, Saudi Arabia reportedly was producing close to this amount. Since that time, however, it has had to progressively cut production in an effort to reduce the world oil glut and stabilize prices. At the end of 1998, Saudi Arabia was producing at levels much closer to 8 million barrels a day. In the three months after the March 23, 1999 OPEC cutbacks effective April 1, 1999, Saudi Arabia produced an average of 7.477 MMBD, slightly over its 7.438 MMBD qouta.

The Development of Saudi Oil Reserves

Although it has averaged over 2 million barrels per day of spare production capacity (the largest in the world), Saudi Arabia has continued to invest in the development of its crude reserves. It is completing the final development of the Shaybah field in the Empty Quarter area bordering the United Arab Emirates. The field contains an estimated 7 billion barrels of 40-42° API sweet crude oil, and will ultimately produce 500,000 barrels per day of crude oil and 870 million cubic feet/day of natural gas. *The field began production in July 1998 at around 250,000 bpd.* The overall project will cost $2-$2.5 billion, and will include three gas/oil separation plants (GOSPs) and a 395-mile pipeline to connect the field to Abqaiq, Saudi Arabia's closest gathering center, for blending with Arabian Extra Light crude. Two U.S. companies are playing a major role in the project: Parsons Corporation (project management) and Bechtel (construction).

Production is being increased in other areas. Japan's Arabian Oil Company (AOC) operates the offshore fields (Khafji, Hout) within the Neutral Zone, with production capacity of about 350,000 barrels per day. Texaco operates the onshore fields (Wafra, South Fawaris, South Umm Gudair), with current crude oil production of more than 200,000 barrels per day. *In* 1998, Texaco *planned* to increase output from its fields and also drill three exploration wells in the Neutral Zone. AOC's concession on the Saudi side expires in 2000 (an extension has been requested), and on the Kuwait side in 2003. Texaco's onshore concession lasts until 2010. In march 1999, Saudi Arabia Petroleum & Mineral Resources Minister Ali Naimi suggested that Japan would have to meet certain conditions, including increased investment in Saudi Arabia and the financing of a railroad, for AOC to renew its concession. Though the Japanese have not been prepared to meet the Kingdom's demands, its proposals are still larger than most put forth by the American majors.

Saudi capacity to increase production is so great that the EIA estimated in 1998 that the long-term growth in Saudi oil production would outstrip the growth in all of the nations in the Former Soviet Union, in spite of major increases in production by the former Soviet republics in the Caspian and Central Asia. The EIA estimated in 1999 that Saudi Arabia would increase its production from 8.6 million barrels per day in 1990, and *11.4* million barrels per day in 1997, to *11.1* (*11.5-11.2*) million barrels per day in 2000, *13.7* (*12.4-13.8*) million barrels per day in 2005, *14.1* (*12.9-17.3*) million barrels per day in 2010, *16.2* (*13.3-21.9*) million barrels per day in 2015, and *20.0* (*17.8-27.4*) million barrels per day in 2020.

To put these estimates in perspective, Saudi production is already equal to all other states in the Gulf combined, and the EIA estimates that it will remain so through 2020. Saudi production is *160%* of all production in the FSU today, and the Department of Energy estimates that it will *fall* to *152%* by 2020. Past EIA estimates indicated that Saudi Arabia's production will shift from *14.7%* of world production in *1997*, to *13.8%* in 2000, *14.8%* in 2010, and then rise to *17.8%* in 2020.

Much depends, however, on future world oil demand and prices, and the health of the Saudi economy. So far, Saudi Arabia continues to invest in maintaining and increasing its oil, gas, and petroleum related industries. It has, however, made increasing cuts in its current expenditures on oil and product, partly because it now has substantial surplus proudction capacity, and has tended to concentrate near term spending on gas. Saudi Arabia has also encouraged private investment efforts, some involving foreign partners.

The strategy Saudi Arabia will pursue in the future is a key issue, as is the way in which the Kingdom will seek to fund its strategy and its resulting success. The current economic crisis and drop in oil prices creates major uncertainties about the future impact of the Gulf on world oil supplies, and the future role of private investment. It also highlights the importance of other uncertainties, *such as* Saudi economic reform, Saudi budgets, and Saudi ability to fund the civil investment and entitlement programs that are critical to Saudi security.

The Impact of a Market Share Strategy

This background on Saudi Arabia's existing plans is critical in order to understand the role that a market share strategy might or might not play. There are a number of possible variations on this strategy, but it essentially calls for Saudi Arabia and other low cost Gulf producers to maximize production at low prices.

Saudi Arabia has *an* advantage in any price war strategy because its older fields are so cheap to produce. Saudi Arabia can produces five different grades of crude oil. When oil prices fall, it caps production of some of its lower grade Arabian Heavy crude, pumping the more valuable Arab Extra Light crude from fields like it's new Shaybah field in the Empty Quarter instead. This not only keeps production costs down, but also produces higher value oil products per barrel when refined. Shaybah, which began production in July 1998, pumps 500,000 barrels of oil a day, and has production costs of less than $2 *per* barrel. Shaybah, which cost nearly $3 billion to develop, has proven reserves of 14.3 billion barrels of oil and an estimated life of 75 years. As Saudi Aramco's Chief Executive Officer Abdullah Jum'ah put it, "We will be here for a long time. When the oil runs out, the field will remain active for another 75 years producing gas."

At the same time, such a strategy means *Saudi Arabia must make* further short to midterm budget cuts and taking the political risks inherent in economic reforms. *Such a* strategy also means potential confrontations with Iran, which is a relatively high cost producer, and with Iraq because it cannot take advantage of its status as a potential high volume, low cost producer for at least half a decade because of sanctions and limited production capacity.

The potential advantage of this strategy is that it would put severe pressure on high cost producers in other regions like the Caspian, U.S., and Western Canada. It would *most likely* freeze production in *the* North Sea and West Africa *in order* to compete and attract investment in increasing their volume of production, although Mexico, Venezuela, Libya, and Algeria *are* also estimated to be low cost producers that would benefit *from* increased production. The Southern Gulf would essentially price oil at a point where the marginal return on investment in other regions would be far less attractive and restructure the entire oil industry.

There are several problems with this strategy. First, it involves a massive sustained gamble in macroeconomics over half a decade that is acutely sensitive to the future marginal impact of advances in exploration and production strategy in other regions. Second, it requires major efforts at true economic reform, plus considerable budget austerity and additional private investment. Third, it means a potential confrontation with both Northern Gulf states.

More generally, the net effect of the entire strategy is to raise Saudi production levels to MMBD in 2003. The EIA already projects that Saudi Arabian production capacity will increase from 8.6 million barrels per day in 1990, and 11.4 million barrels per day in 1997, to 11.1 (11.5-11.2) million barrels per day in 2000, 13.7 (12.4-14.8) million barrels per day in 2005, 14.1 (12.9-17.3) million barrels per day in 2010, 16.2 (13.3-21.9) million barrels per day in 2015, and 20.0 (17.8-27.4) million barrels per day in 2020.

This brings up a key issue in terms of mid to long-term energy strategy. Virtually any major problem in terms of energy supply tends to increase the central importance of Saudi Arabia, regardless of whether it pursues a high risk "market share" strategy or not. Saudi Arabia's combination of massive reserves, high existing production capacity, and low production costs means that it will benefit over time from any major depletion of reserves in other countries that raises oil prices, production problems caused by political or economic reasons, slow-downs in technology gain, declines in the finding of new reserves, or major increases in economic growth in any region. Only major new discoveries of alternative fuels or new technologies can prevent Saudi Arabia from receiving some combination of these benefits over time.

References

Marcel, Valerie. 2006. *Oil Titans: National Oil Companies in the Middle East*. London: Chatham House.

Mommer, B. 2002. *Global Oil and the Nation State*. Oxford: Oxford University Press.

Noreng, Oystein. 2002. *Crude Power: Politics and the Oil Market*. London: I.B. Tauris

Paarlberg, Robert. 1978. 'Food, Oil, and Coercive Resource Power.' *International Security*, Vol. 3, No. 2 (Autumn), pp. 3-19.

Russett, Bruce M. 1993. *Grasping the Democratic Peace: Principles for a post-Cold War World*. Princeton: Princeton University Press, 1993.

Bibliography

Ammann, Daniel. 2009. *The King of Oil: The Secret Lives of Marc Rich*. New York: St. Martin's Press.

Assis, Claudia, 201. Oil futures extend gains amid unrest, *USA Today,* 23 February.

Blank, Stephen. 1995. 'Energy, Economics and Security in Central Asia: Russia and its Rivals.' *Central Asian Survey*, 14 (3), pp. 373–406.

Bolinger, M., Wiser, R., Milford, L., Stoddard, M., and Porter, K. 2001. States emerge as clean energy investors: A review of state support for renewable energy. *The Electricity Journal* (Nov.), 82-95.

Carlsson, B.; R. Stankiewicz. 1991. "On the nature, function, and composition of technological systems". *Journal of Evolutionary Economics* 1: 93–118.

Carr, E. H. 1946. *The Twenty Tear's Crisis: an introduction to the study of International Relations*, London: Macmillan.

CIA, World Factbook, 1997, Washington, GPO, CD ROM, "Saudi Arabia;" The World Bank, World Bank Atlas,1998, Washington, World Bank, pp. 36-37.

Dunn, S. 2000. *Micropower: The Next Electrical Era.* Worldwatch Institute, Washington, DC.

Energy Information Agency, 1999, *International Energy Outlook*, Washington, DOE/EIA-0484(99), March 1999, p. 32.

Freeman, C. and F. Louca. 2002. *As Time Goes by: From the Industrial Revolutions to the Information Revolution.* Oxford: Oxford University Press.

Friedman, Thomas. L. 2006. 'The First Law of PetroPolitics' *Foreign Policy*, May/June 2006.

Frynas, Jedrzej George. 2009. *Beyond Corporate Social Responsibility. Oil Multinationals and Social Challenges*. Cambridge: Cambridge University Press.

Gilpin, Robert. 1987. *The Political Economy of International Relations.* Princeton: Princeton University Press

Hinnebusch, Ray. 2003. *The International Politics of the Middle East.* Manchester: Manchester University Press.

Hirsh, R.F., and Serchuk, A.H. 1999. Power switch: Will the restructured electric utility system help the environment? *Environment 41*, 4-9; 32-39.

IEA (International Energy Agency). 2000. *World Energy Outlook 2000*. Organization for Economic Cooperation and Development (OECD)/IEA

International Energy Agency (IEA). 2009. *Key World Energy Statistics 2009*. Printed in Paris, France by Soregraph.

Jacobsson, S.; A. Bergek. 2004. "Transforming the energy sector: the evolution of technological systems in renewable energy technology". *Industrial and Corporate Change* 13 (5): 815–849.

Jafar, Majid. 2004. 'Kazakhstan: Oil, Politics and the New 'Great Game'. In Shirin Akiner (ed.), *The Caspian: Politics, Energy and Security*. London: RoutledgeCurzon.

Kaldor, Mary; Terry Lynn Karl and Yahia Said. 2007. *Oil Wars*. London: Pluto Press.

Kandiyoti, Rafael. 2007. *Pipelines: Flowing Oil and Crude Politics*. London: I.B. Tauris

Karasac, Hasene. 2002. 'Actors of the New 'Great Game.' Caspian Oil Politics.' *Journal of Southern Europe and the Balkans*, 4 (1).

Klare, Michael. 2008. *Rising Powers, Shrinking Planet: The New Geopolitics of Energy*. New York: Henry Holt and Company.

Madubansi, M. & Shackleton, C. M. 2006. Changing energy profiles and consumption patterns following electrification in five rural villages, South Africa. *Energy Policy*, 34 (18): 4081-4092

Marcel, Valerie. 2006. *Oil Titans: National Oil Companies in the Middle East*. London: Chatham House.

Marsh, R., Larsen, V. G. & Kragh, M. 2010. Housing and energy in Denmark: past, present and future challenges. *Building Research & Information*, 38(1): 92-106.

Martinot, E., Chaurey, A., Moreira, J., Lew, D., Wamukonya, N. 2002. Renewable energy markets in developing countries. *Annual Review of Energy and the Environment 27*: 309-348.

McKinsey Global Institute (MGI). 2007. Curbing Global Energy Demand Growth: The Energy Productivity Opportunity. *MGI publication.*

Mommer, B. 2002. *Global Oil and the Nation State*. Oxford: Oxford University Press.

Noreng, Oystein. 2002. *Crude Power: Politics and the Oil Market*. London: I.B. Tauris

Paarlberg, Robert. 1978. 'Food, Oil, and Coercive Resource Power.' *International Security*, Vol. 3, No. 2 (Autumn), pp. 3-19.

Petroleum Finance Corporation, 1998, "Saudi Oil Policy Options – Time to Choose: Market Share or Price Defense?,"Washington, December.

Russett, Bruce M. 1993. *Grasping the Democratic Peace: Principles for a post-Cold War World*. Princeton: Princeton University Press, 1993.

Soares de Oliveira, Ricardo. 2007. *Oil and Politics in the Gulf of Guinea*. New York: Columbia University Press.

Sprout, Harold and Margaret Sprout. 1971. *Towards a Politics of Planet Earth*. New York: Van Nostrand Reinhold

Therramus, Tom. 2010. "Oil Caused Recession, Not Wall Street". Oil-Price.net. Retrieved 21 March 2012.

Turner, Louis. 1978. *Oil Companies in the International System*. London: Royal Institute of International Affairs

U.S. Arms Control and Disarmament Agency (ACDA), 1996, *World Military Expenditures and Arms Transfer*, GPO, Washington.

Watts, Michael. 2009. 'Crude Politics: Life and Death on the Nigerian Oil Fields'. *Niger Delta Economies of Violence Working Papers*. Institute of International Studies, University of California.

Wenger, Andreas; Robert W. Orttung, Jeronim Perovic (eds). 2009. *Energy and the Transformation of International Relations*. Oxford: Oxford University Press.

William Engdahl F. 2012. "Behind Oil Price Rise: Peak Oil or Wall Street Speculation?". *Axis of Logic*. Retrieved 21 March 2012.

World Bank, 1998, World Development Indicators, Washington, World Bank, p. 14.

Worthington, R. K. 1984. "Renewable Energy Policy and Politics: The case of the windfall profits tax". *Policy Studies Journal*: 365–375.